YOU BE THE JUDGE . . .

- **Gulf Breeze, Florida.** In 1987 hundreds of UFO sighting reports came in from this small coastal community, and resident Ed Walters at last felt he had to tell his own shocking story.
- **Betty and Barney Hill.** Probably the most famous abduction on record, this 1961 New Hampshire incident brought agent Pope back for another look, and his own conclusion.
- **Alan Godfrey.** A credible, solid British constable, Godfrey's tale of a UFO encounter and his abduction emerged under hypnosis, and his experience startled a nation.
- **Whitley Strieber.** This writer's bestseller *Communion* brilliantly described his encounters with "visitors." The book earned a fortune, but was it fact or a figment of a vivid imagination?

READ THE REPORTS. LISTEN TO THE FACTS.
THEN DISCOVER A GOVERNMENT OFFICIAL'S
STUNNING CONCLUSIONS ABOUT . . .

THE UNINVITED

ALSO BY NICK POPE

Open Skies, Closed Minds
Official Reactions to the UFO Phenomenon

THE UNINVITED

An Exposé of the Alien Abduction Phenomenon

Nick Pope

A Dell Book

Published by
Dell Publishing
a division of
Random House Inc.
1540 Broadway
New York, New York 10036

Copyright © 1997 by Nick Pope

The trademark Dell® is registered in the U.S. Patent and Trademark Office.

ISBN: 0-440-23487-5

Designed by Sabrina Bowers

Reprinted by arrangement with The Overlook Press

Printed in the United States of America

Published simultaneously in Canada

April 1999

10 9 8 7 6 5 4 3 2 1
OPM

**To Dad and Helen
With Love**

ACKNOWLEDGMENTS

It's a cliché, but a truism no less, that the author of any book is just one of a number of people who helps bring it to life. I'd like to pay a heartfelt tribute to some of the wonderful people who have played their part in this project.

I'd like to think my agent, Andrew Lownie, for his tireless efforts on my behalf, and for his encouragement over the last three years. His energy and determination have been invaluable to me in my writing career, which would never have got off the ground without his help.

As ever, the entire Simon & Schuster team have been fantastic, and have given me every support and encouragement. I'm especially grateful to Martin Fletcher, whose enthusiasm for this book has been inspirational, and whose input has helped shape the entire project. I'm enormously grateful to Sally Partington, whose constructive copy-editing considerably enhanced the text. Lisa Shakespeare, as ever, has arranged a phenomenal range of media activities, and has given me the most wonderful support throughout. I'd also like to thank Cathy Schofield, Keith Barnes, Gillian Holmes and Glen Saville.

I'm grateful to those abduction researchers, ufologists, writers and witnesses who have helped me with material, contacts, good advice or friendship. I'd particularly like to thank Timothy Good, John Spencer, Jenny Randles, Peter Robbins, Betty Hill, Whitley and Anne Strieber, David Pritchard, Roy Lake, Tony Dodd, Philip Mantle and Harry Harris.

I'd also like to pay tribute to those abduction researchers whose work has given me inspiration, and whose pioneering efforts gave me the courage to look beyond my initial skepticism. There are far too many to mention individually, but I must single out Jacques Vallée, David Jacobs and John Mack, who have pointed the way for so many.

I don't think it's possible to research the subject of alien abductions without acknowledging the enormous contribution made by Budd Hopkins. I regard him as being the founding father of modern abduction research, and a man without whom we would all be very much more in the dark than we are now. If one person deserves to crack the mystery because of his dedication and compassion, it's Budd.

I'm incredibly grateful to those abductees, experiencers, participants and witnesses who have allowed me into their lives, enabling me to get a better insight into what is going on. Their courage and dignity in the face of the unknown is extraordinary. I'd like to thank Patsy, Mary, Peter, Jayne, Maria, Vaunda and Chris for allowing their stories to be featured here. Some of the experiences have been traumatic, but those concerned have placed their own feelings on one side, and are driven by a desire to help and encourage others who have experienced this phenomenon. I'd also like to thank and pay tribute to the many others whom I've interviewed, but whose stories I haven't featured. Thank you for your help and your inspiration.

I've had a lot of encouragement from a number of friends and colleagues at the Ministry of Defence—civil servants, military and other specialists. Some researchers will doubtless read much into such words, and formulate theories about cover-ups, conspiracies or the controlled release of information. This would be a mistake. It is simply that there are some at the Ministry who genuinely feel that there is good evidence for alien abductions and UFOs, and who want more official action. Opinions *are* slowly changing, but these are still difficult subjects to be too closely associated with, and for this reason I shall not mention any names. Those concerned know who they are, and know how much their support has meant to me.

I am also grateful to A. P. Watt Ltd, who on behalf of the Literary Executors of the Estate of H. G. Wells gave me permission to quote from *The War of the Worlds*, still a masterpiece, even one hundred years after its first appearance.

On a personal level I'd like to thank my Dad and my stepmother, Helen, for their love, support, encouragement, help and words of wisdom. I'd also like to thank my brother Seb, Laura, Jared, Lizzie, and all my other friends who have had to put up with my obsession over the last year or so, as I locked myself away to write this book.

A word also for those people whom I have inevitably forgotten to mention. I am no less grateful to them, and hope they will forgive the oversight.

Last, but certainly not least, I'd like to thank Michèle Kaczynski, whose support and encouragement has been priceless. She has kept me on schedule with my writing, offered valuable suggestions on the text, helped me with various stages of proofreading, taken on numerous administrative tasks and done much more besides. I am grateful beyond words.

We have learned . . . that we cannot regard this planet as being fenced in and a secure abiding place . . . we can never anticipate the unseen good or evil that may come upon us suddenly out of space.

H. G. Wells
The War of the Worlds

INTRODUCTION

All around the world, quite independently, thousands of people claim that they have been abducted by aliens. At first sight these claims seem ludicrous, but when one takes a closer look, it soon becomes clear that they are supported by some very convincing evidence. Clearly something strange is going on. But what, exactly?

I spent three years of my life in a job where my duties included investigating UFO sightings for the Ministry of Defence. Between 1991 and 1994 my duties at the Ministry involved me in a search for the truth about one of the most powerful and enduring mysteries of modern times.

My first book, *Open Skies, Closed Minds,* told the story of my three-year voyage of discovery, and detailed some of what television devotees might like to think of as "the real *X-Files.*" I had come into the job as a skeptic, but emerged believing that a small percentage of UFO sightings did involve extraterrestrial craft. My conversion was not a blind leap of faith, but was based upon numerous instances where my rigorous official investigations had failed to uncover any conventional explanation for what was seen. These cases included visual sightings backed up by radar evidence, and UFO reports from civil and military pilots. In 1980 there was a case where radiation readings ten times normal were recorded on the spot where a metallic craft landed near two military bases (the famous Rendlesham Forest case). In 1990 the Belgian Air Force scrambled F-16 fighters to try and intercept a UFO that had been picked up on radar by various NATO and Belgian installations, but the UFO easily evaded the jets. In 1993 a craft flew directly over two Royal Air Force bases in England, firing a beam of light at the surrounding countryside. There were plenty of cases where the intruding technology seemed to be considerably more advanced than the defending technology.

Fascinating and disturbing though some of these UFO incidents were, it was always the reports of abductions that gave me the greatest concern. The UFO mystery paled into insignificance when compared to the abduction phenomenon. Although there will be much talk about extraterrestrials, I should make it clear that this book is not about UFOs. Although, of course, UFOs will be featured, the central issue here is encounters between humans and other, non-human intelligences. Using an analogy to illustrate the point, if the Queen called at your house, you would probably not be that interested in the car in which she arrived. You would want to know why she was visiting you! In the same way, for those like me who believe that some UFOs are extraterrestrial in origin, the UFOs become, simply, a means of transportation. As the respected ufologist Stanton Friedman has said:

"Never mind the saucers; what about the occupants?"

For me, carrying out my job at the Ministry, the whole issue of alien abductions raised serious defense and national security issues. If the accounts were to be believed, then not only were UFOs penetrating our sophisticated air defense network with impunity, but the occupants of the craft were sometimes carrying out intrusive and frightening procedures on unwilling human subjects.

When I first encountered references to alien abductions, I was skeptical. Similarly, when I was actually introduced to people who claimed that they had been abducted, I found myself looking for other, more conventional explanations. Were they simply lying, or were they fantasy-prone personalities, who might construct an abduction story for psychological reasons? Might there be mental health explanations, with some sort of psychopathology lying at the root of these people's claims? Could it be that certain stimuli had combined to create mass hysteria on a frightening scale? All these options needed to be explored, and would in themselves have been worthy of serious study. But the most frightening theory of all was that the claims were true, and the events were occurring as described. If they were true—if just *one* case was true—the implications for the human race would be profound and disturbing.

It is clear that, despite the unwelcome intrusion of various cranks, most of the so-called abductees who have come forward to talk about their experiences are sincere and well-intentioned. For this reason, although I believe that many such experiences *do* have conventional explanations, I shall not be inserting the word "alleged" before every mention of an encounter. Whatever the cause, these experiences are real, in the sense that they are perceived as such by the majority of those involved. The key issue, of course, is whether or not any of these encounters take place in the physical universe, as opposed to the psychological one.

A fundamental issue that I shall be addressing is whether the experiences are positive or negative. This may at first sight seem a ridiculous question. How can the taking of someone against their will ever be anything other than negative? There are, however, many people who see the experience as something wonderful, and a whole New Age religion has grown out of reports of encounters with extraterrestrials. And as we shall see, some abductees have benefited from their experiences. Consequently, many people argue that it is unfair to use words like "aliens" when describing the extraterrestrials. They say this introduces an automatic negative bias, because so much of society regards with fear and suspicion anything that is foreign or unknown. It's a fair point, although I can't help thinking that the New Age preference for the term "visitors" takes political correctness just a little too far! Similarly, many of those with a New Age view of this phenomenon talk not of abductees, but of "experiencers." I have generally avoided such terms in this book. Not because I am unreceptive to New Age ideas, but because I believe phrases like "alien abduction" state the issue plainly, and focus attention more clearly on the central problem that we face when trying to evaluate this terrifying and elusive phenomenon. It is important that we bear this terror in mind, because it is a central feature of many of the cases. Talk of "visitors" and "experiencers" seems to me to downplay the suffering that is undoubtedly involved in many cases.

Although the primary aim of this book is to give an overview of the alien abduction phenomenon, especially for those who are coming to the subject for the first time, I shall be covering in depth a number of entirely new cases that I have investigated. By combining history, theory and casework, I aim to present as clear a picture as possible of a phenomenon that lies at the cutting edge of science, and at the boundary of reality itself.

CONTENTS

PART 1

THE PHENOMENON

CHAPTER 1

The Little People

How can we begin to define the alien abduction phenomenon when ufologists are unable to reach any consensus on the subject, while conventional science refuses to accept it at all? We do, however, need to start somewhere, and it is worth looking at the definition of an abductee that has been drawn up by the leading American UFO group, the Center for UFO Studies (CUFOS). They define an abductee as follows:

> *A person who was taken against his or her will from terrestrial surroundings by non-human beings. The*

beings must take the person to an enclosed space not terrestrial in appearance which the person assumes or knows to be a spacecraft. In this place the person must either be subjected to an examination, engage in communication (verbal or telepathic) or both. These experiences may be remembered consciously or through methods of focused concentration (e.g. hypnosis).

Although this is probably as good a definition as any that I have seen, it seems to me that any formal criteria are self-evidently limiting. Such rules are likely to exclude experiences that fail to fit in with the belief system of the investigator, even though the experiences themselves may be real. Central to this is the issue of consent. Does an abduction occur only when somebody is taken against their will? What of someone who gives consent out of fear or curiosity? What if someone is tricked into participating in an experience? For these reasons, I prefer to talk about a phenomenon that I define as "Human encounters with non-human intelligences." This rather wider definition avoids the limiting criteria of the CUFOS definition, and also ensures that I am not tied to the acceptance that the abductors are necessarily extraterrestrials. Far more importantly, it removes the requirement that the experience involves a spacecraft. This is crucial, because the concept of a spacecraft has existed only for a century or so, starting with the science fiction works of Jules Verne and H. G. Wells, while human encounters with non-human intelligences have been going on, I believe, throughout history.

In other words, any definition of abduction that is couched in modern terms probably excludes ninety-nine percent of the phenomenon.

While it is tempting to think of the alien abduction phenomenon as something very new, I believe the opposite to be the case. Far from being the recent, American phenomenon that many believe it to be, it is my view that such events have been occurring for as long as humans have inhabited this planet. Although the frequency of these occurrences may be increasing, and although the phenomenon itself may be evolving, if we are to place abductions in their proper historical context, we need to go back in time. But in doing so, we must tread outside the bounds of our recorded history. History is biased. It is inevitably written by the victors of conflicts, or by those who seek to perpetuate their own social, economic, political or religious views. History has also tended to be written by members of the ruling classes within any society, largely because the working classes have so seldom had the educational opportunities that might have given them the literacy required to write their own histories.

There is, however, a body of recorded history that is in many ways more honest than the more formal accounts that we learn, parrot fashion, at school. There is a way in which information has been handed on, orally, from generation to generation over the years, in a way that bypassed the control of the ruling oligarchies. We call this *folklore*.

Folklore consists of the traditional beliefs and tales of communities; is found in all cultures of the world; and is not dependent upon literacy. A key point about folklore is that while it is so frequently told as fiction, it often has its roots in fact.

One of the most common and enduring themes running throughout folklore is that of encounters between humans and non-human intelligences. So at the beginning of our quest for the truth about alien abductions, we need to return to the folklore, and especially to accounts of the little people, who have always been key players in folkloric history.

Fairy, dwarf, goblin, hobgoblin, pixie, elf, sprite, abatwa, sidhe. These are just a few of the words used to describe the little people. These accounts come from all around the world, and one of the most likely explanations when an account of something is found in such a cross-cultural pattern is that what is being described genuinely exists, in some form. In other words, the little people are real. Modern science may scoff at such ideas, but contemporary, industrialized, Western society is just about the only culture in human history not to have what I call a corporate belief in such things. And even within this agnostic culture, one only has to scratch the surface to find the old beliefs alive and well. Go beyond the cities, and out into the rural communities, and you will soon find places in which these beliefs still thrive. Celtic folklore is particularly rich, and in Ireland, belief in the little people (sometimes known as The Gentry) is widespread, even to this day. An Irish friend of mine has told me how during her childhood it was common practice to leave food out for the fairies.

Much of the research into the links between folkloric

encounters with the little people and modern encounters with aliens has been carried out by the French astrophysicist Jacques Vallée, whose book *Dimensions* draws together a wealth of historical reports from cultures all over the world. It is readily apparent that there are huge similarities between the old and the new accounts; the only thing that changes is the language. Vallée believes that these creatures—whatever name we give them—are fellow inhabitants of what he calls the *Multiverse*. As we shall see, this is but one of many theories about these *other* intelligences.

Further work on the relationship between folklore and abduction reports has been done by the American scholar Thomas E. Bullard, who holds a Ph.D. in folklore. Bullard has argued that where folklore derives purely from the imagination, there is a huge variety in the tales told, as each storyteller adds their own embellishments. But Bullard draws a distinction with modern abduction reports, which are, as we shall see, remarkably similar to one another. He believes this uniformity in abduction accounts differentiates them from the purely fictitious folklore.

So if we apply this principle to the various disparate myths about the little people, perhaps we can identify the common features, and thus isolate the core phenomenon.

The most significant physical characteristic is of course the small size of many of these strange beings. Straight away one sees a link between these folkloric descriptions and more modern accounts of the small, gray beings often reported as being responsible for alien abductions. We will be meeting these so-called Greys later, but first we must familiarize ourselves with their precursors.

A theme that runs through many accounts of the little people is their interaction with humans, and this is particularly true of the fairy folk. Such entities are commonly described as being highly intelligent, and prone to taking mortals away to fairyland. Most fairy activity occurs at night, and often includes abducting a human baby and replacing it with a replica, known as a *changeling*. Fairies from Welsh mythology—known as the Tylwyth Teg—are famed for this, and children taken by the Tylwyth Teg, if recovered, will have only vague memories of their experiences. (As we shall see, this amnesia is another key feature in the modern abduction accounts.) These cross-cultural folkloric beliefs about the taking of babies, and about changelings, are mirrored in accounts of alien abductions where the creation of hybrid human/alien babies is a central theme, and is the favorite theory of those ufologists who make a particular study of abductions. Variations on this theme include the accounts of marriages between fairy folk and humans, and intriguing stories of midwives being abducted to assist with fairy births. Again, there is an inescapable parallel with the whole concept of the genetic breeding program that many ufologists insist lies behind the modern abduction phenomenon. Have humans been participants—sometimes willing, sometimes not—in an age-old but still continuing program to unite two intelligent races?

There are other common themes; other clues that point to a link between the ancient and the modern accounts. In fairyland, time does not run as it does in the normal world. Witness the many accounts of people taken away by the fairies who find that on their return much more time has elapsed than they had realized. The best known such tale is Washington Irving's *Rip Van Winkle*, where the eponymous hero encounters strange folk while out walking in the hills. He drinks from a large keg, falls into a deep sleep, and awakes to find his surroundings have changed. When he returns to his village, nobody recognizes him, and it transpires that he has slept for twenty years. This, of course, is fiction, but Irving was well aware of much more ancient stories of people having fallen into a deep sleep for many years, or having been taken away for some years to fairyland. There are similar legends to *Rip Van Winkle* in Welsh, Teutonic and Chinese folklore. Do such tales have their roots in reality? Could they be early—and somewhat exaggerated—accounts of the so-called *missing time* phenomenon, which modern ufologists see as being the primary clue that an abduction experience has taken place? In these accounts a traveler arrives home only to find that the time is much later than she would have expected. Here, a forgotten experience (i.e. an abduction) is believed to have taken place during the lost hours.

Folkloric tales of human interaction with fairies often focus on the otherworldly nature of such encounters. It is not simply that time behaves in peculiar ways; there is an altered state of consciousness. The participants seem to enter another universe altogether. This is an effect that is also to be found in many modern-day alien abductions. Typically, witnesses will comment that all normal background noises and activity stopped. If they are driving cars, they may notice that theirs is the only car on the road, even if the event occurs in a busy area. This effect is sometimes

known as *The Oz Factor*, because of the analogy with
Dorothy being taken to a magical world in *The Wizard of
Oz*. British ufologist Jenny Randles, who has made a particular study of this, has noted that the Oz Factor also occurs in paranormal events such as time slips.

Readers will have noted that I differentiate between
abductions and paranormal phenomena. I will discuss
later the possibility of a psychic element to alien abductions, but it seems to me that we must first explore the
possibility that these events occur within the physical universe. After all, if UFOs are extraterrestrial spacecraft,
then their visits here, and the abductions, may be likened
to the first contact between Europeans and Native Americans. The European ships were like nothing the Americans
had previously encountered, and the Europeans, though
similar in appearance to the indigenous people, were not
identical. The whole situation may have been extremely
strange to the Native Americans, but in no sense was it
paranormal.

One legend that links the idea of time behaving
strangely with the concept of a mysterious *otherworld* is
that of the Irish warrior-hero Oisín, who fell in love with a
fairy maiden, and went to live with her in Tir-Na-N'Og
(the land of the young). He is blissfully happy there, but
one day decides he wants to visit an old friend. He is given
a white horse, but told he must not dismount while back in
the mortal world, or he can never return. In the mortal
world he sees that despite believing he had been away for
only a short while, years have elapsed, and his friend has
been long dead, while he has remained young. At this
point he dismounts, to help some men lift a stone, and can

thus never return to Tir-Na-N'Og. In another Celtic tale, a boy who is lost in a forest is scared by an owl, and then hears strange whispering sounds. He catches sight of a bright, white ball of light through the trees, which he follows. He is led to a beach, where a magical white horse transports him to Tir-Na-N'Og.

The little folk are notorious shape-shifters, often able to disguise themselves as various animals, and sometimes able to make themselves invisible. This is another concept that we will encounter when we look at more modern accounts of alien abductions, together with that of a screen memory—where the abductee somehow perceives the alien not in its true form, but as a terrestrial creature such as a deer, a cat, or an owl. Although outside the scope of this book, it is perhaps worthy of note that more malevolent shape-shifters are a common theme in world folklore, and the idea of vampires and werewolves is deeply rooted in many different cultures. David Farrant, who investigates sightings of vampires, has a number of cases where strange figures with striking eyes have visited people's bedrooms at night, in situations where the witnesses have been temporarily paralyzed. As we shall see, this is a common scenario in the alien abduction phenomenon.

Much of our rich legacy of folklore has been hijacked by the Church. Early Christian authorities feared competition from pagan religions and mythologies, but were sophisticated enough to realize that the best way to deal with this competition was not to attempt to suppress the folklore but to incorporate it into their own religious structures. In this way, for example, the festival of Christmas simply supplanted existing pagan ceremonies that had traditionally taken place at around the time of the Winter Solstice. And as far as tales of the little people were concerned, the figure of the Blessed Virgin Mary and the saints were similarly incorporated into many existing stories. Ironically perhaps, the theme of UFOs and abductions runs through the Bible, which abounds with accounts of people being taken up into the sky. The phrase "The chariot of fire" comes straight from the Bible, and demonstrates how any unknown phenomenon can be described only in the language available to the teller of the tale. Jacob's dream of angels ascending and descending the ladder between earth and heaven is, perhaps, another interesting biblical parallel with our idea of aliens and UFOs.

There is certainly one biblical account that could almost read like a modern UFO report, complete with the sighting of entities. This occurs in the Book of Ezekiel, where the prophet makes the following observation:

> And I looked, and, behold, a whirlwind came out of the north, a great cloud, and a fire infolding itself, and a brightness was about it, and out of the midst thereof as the color of amber, out of the midst of the fire. Also out of the midst thereof came the likeness of four living creatures. And this was their appearance; they had the likeness of a man. And every one had four faces, and every one had four wings. And their feet were straight feet; and the sole of their feet was like the sole of a calf's foot: and they sparkled like the color of burnished brass . . . As for the likeness of the living creatures, their appearance was like burning

coals of fire, and like the appearance of lamps: it went up and down among the living creatures; and the fire was bright, and out of the fire went forth lightning . . . Now as I beheld the living creatures, behold one wheel upon the earth by the living creatures, with his four faces. The appearance of the wheels and their work was like unto the color of a beryl; and they four had one likeness: and their appearance and their work was as it were a wheel in the middle of a wheel. When they went, they went upon their four sides: and they turned not when they went. As for their rims, they were so high that they were dreadful; and their rims were full of eyes round about them four. And when the living creatures went, the wheels went by them: and when the living creatures were lifted up from the earth, the wheels were lifted up . . . And when they went, I heard the noise of their wings . . .

Does this description not sound remarkably like an encounter with four non-human intelligences, who have landed and then taken off in a UFO?

Related to these biblical accounts are stories about the sky people, sometimes also known as the star people. The idea of heaven being in the sky is not one that originated with Christianity. Native American culture is particularly full of tales about sky people, and tribes such as the Iroquois believe that humans actually originated from the sky. Another common Native American myth was that certain exalted humans were taken up into the sky, where they become constellations. Is it a coincidence that almost identical legends were prevalent half a world away, in the culture of ancient Egypt, which was in turn adopted by the Greeks and the Romans? Could there be a link, and could it be that the tales of mortals being taken up into the sky were literal, and not simply allegorical?

So what might we deduce if we place all these accounts together? When we weigh up all the stories, whether they be biblical or folkloric, we do find common threads. We see stories of people living in the sky, and coming down to the ground. We have numerous accounts of the little people, who are variously described as intelligent, mischievous, and fond of taking humans to fairyland. We have the concept of time running at a different speed from normal, and of a kind of otherwordly state where everything is not quite as it seems.

Skeptics will argue that it is impossible to separate fact from fiction when looking at religious texts and folklore. It is certainly true that biblical stories were often allegorical, and that successive translations have altered the meaning of various passages over the years. Similarly, nobody would maintain for a moment that folklore represents a true account of actual events (but then neither do our biased and selective history books). But many of these accounts have their roots in fact, and there have undeniably been thousands of sightings of the little people over the years.

I believe that all these varied accounts stem not from some mass hysteria, but from the simple fact that people really *have* been interacting with otherworldly beings since the dawn of time. So it may be that what we now call the alien abduction phenomenon is something that has been with us always, perceived in different forms according to our culture and our belief systems. As Ralph Noyes, former Ministry of Defence Under Secretary, and current Honorary Secretary of the Society for Psychical Research has said:

> *We used to populate the Earth with spirits and Gods. Now they have been chased away and the sky is their haven.*

To illustrate the importance of perceptions in defining an experience, consider the following example. If a small gray figure walked through a ufologist's bedroom wall and stood at the foot of the bed, the ufologist would doubtless believe that he or she was having an alien encounter, and was about to be abducted. But suppose this same experience was witnessed by people with no knowledge of UFOs and aliens, but an intense interest in ghosts? They would probably say they had seen a ghost. Christians might believe they had seen an angel. But these are just labels. Perhaps we see what we expect to see, or sometimes what we *want* to see. The experience may occur in the physical universe, but it will inevitably be interpreted psychologically, on a very individual basis, using reference points from the individual's own knowledge, experience and belief.

If I am right, and if ancient encounters with the little people reflect the same experience as modern alien abductions, then are ufologists now correctly interpreting a phenomenon that has been misinterpreted for centuries? Not necessarily. Why should the idea of space aliens be any more or less valid than older theories about fairies? It is possible that this modern view is equally wrong, and that the truth is something so bizarre and so abstract that we can only guess at it. Perhaps the core experience is so strange that it will *always* lie beyond human comprehension.

CHAPTER 2

The Space People

We have seen that the concept of human interaction with other, non-human intelligences is age-old. There is a bridge between the ancient and modern worlds that otherworldly entities traverse. As we move forward in time, leaving the folklore behind us, we approach the world of the abductees. But before we meet the abductees, we have to meet their immediate predecessors, the contactees, who claimed to have encountered and conversed quite openly with benevolent "space people" who were generally humanoid in appearance. And before we meet the contactees, we need to understand that they emerged in

America, against the background of a national obsession with UFOs that started fifty years ago, and is getting stronger by the day.

Setting aside the historical sightings of celestial chariots and cloudships, the modern UFO story really started on June 24, 1947 with a civilian pilot named Kenneth Arnold. Arnold was flying over the Cascade Mountains in Washington State, looking for the wreckage of a C-46 military transport aircraft that had crashed in the area. He didn't find the aircraft, but what he saw changed his life—and the world—forever. He witnessed nine strange craft, flying in formation, and estimated that they were traveling at speeds of about 1,500 m.p.h. This was considerably faster than the top speed of any aircraft in the world at that time. He said that the craft moved in a highly unusual jerking manner, totally unlike an aircraft. When asked to elaborate, he explained that they moved, "like a saucer would if you skipped it across the water." The media coined the phrase "flying saucer," and so began a modern mystery. In fact, Arnold had described the objects as being shaped like a boomerang, and the saucer analogy applied to their movement, not their appearance. But the media were not going to let the facts get in the way, and the public was misled into thinking that Arnold saw saucer-shaped craft, when the truth was he saw no such thing. Other UFO reports began to come in, but none more remarkable than one from a key American military base, which in 1947 was the home of the only atomic-bomb-capable squadron anywhere in the world.

In July 1947 a press release was issued from Roswell Army Air Field in New Mexico that stated:

The many rumors regarding the Flying Disc became a reality yesterday when the Intelligence Office of the 509th Bomb Group of the Eighth Air Force, Roswell Army Air Field, was fortunate enough to gain possession of a disc . . .

Within twenty-four hours of this press release, a cover story about a weather balloon had been deployed. This was clearly ridiculous; the Intelligence Officer of the 509th Bomb Group, Major Jesse Marcel, had witnessed the strange wreckage, and would have known a weather balloon when he saw one, not least because he frequently saw such balloons. Despite the cover stories, it was clear that the US military was convinced, at least privately, that something very strange was going on.

Flying saucer fever had hit America, and this new mystery was soon being talked about all over the world. Although the crash of a disc at Roswell had been effectively covered up, sightings of objects came in thick and fast. By September 1947 the United States Air Force set up its first official study into the situation. Known as Project Sign, this ran until 1949, when it became Project Grudge, before evolving into Project Blue Book in 1951. Blue Book ran for eighteen years, before being closed down in 1969, ostensibly because no evidence of any threat had been found in all the years of study. Ufologists correctly point out that you can't be certain that the UFO phenomenon poses no threat until you know what the objects are, and interestingly, despite the closing down of Project Blue Book, some twenty-three percent of its sightings remained unidentified.

This then was the background to the appearance of the contactees. But we should remember the sociological as well as the ufological context. The world had just emerged from one of the most sickening conflicts in human history. The exposure of the truth about the Holocaust had set a new benchmark for evil, and the start of the Cold War, coupled with the memory of Hiroshima and Nagasaki, had focused people's attention on a harsh new reality. A message of hope was needed. When it came, it was from a totally unexpected source.

George Adamski

George Adamski was born in Poland in 1891 but when he was two, his parents emigrated to America and settled in New York. In 1913 he joined the United States Army, and served with the 13th Cavalry until obtaining an honorable discharge in 1919. Adamski was certainly a complex character, and could perhaps have been described as one of the first of the New Age Californians when, in his late thirties, he drifted into Laguna Beach. His interests were esoteric; he called himself Professor of Oriental Mystical Philosophy, although it is not entirely clear where this title came from, or whether it was one that he had legitimately earned. A precursor of today's cult leaders, his Royal Order of Tibet, which set up a farm at Valley Center in 1940, might legitimately be regarded as a forerunner of the hippie communes that were to spring up some years later. In many ways Adamski was a man born ahead of his time.

Although it is commonly believed that Adamski held a Professorship at the Mount Palomar Observatory, the truth was that he worked at a hamburger stand in the vicinity of the Observatory. His habit of exaggeration, both of his qualifications and his career, has done much to undermine his credibility.

In 1953 Adamski went public with his story, which involved several contacts with extraterrestrial visitors. These claims were made in a book entitled *Flying Saucers Have Landed*, which Adamski had co-written with a British author, Desmond Leslie. The book became a bestseller and, over the next few years, Adamski went on to become well-known all around the world. He met Queen Juliana of the Netherlands, and claimed to have had a meeting with the Pope—although this is something that the Vatican denies. I have been told by someone who was present that a lecture given by Adamski in London in 1963 was attended by Air Chief Marshal Lord Dowding—who had been Commander-in-Chief of Fighter Command during the Battle of Britain—and Lord Mountbatten of Burma.

According to Adamski, he had been having UFO sightings since 1946 (placing his first sighting a year before Kenneth Arnold's famous 1947 encounter). But his landmark contact had apparently occurred on November 20, 1952, after he and six other people drove into the desert to see whether they would be able to spot and perhaps photograph a UFO. Before long they sighted a cigar-shaped UFO, and Adamski took a number of photographs with a camera attached to the end of his telescope, which in turn was mounted on a tripod. Adamski then went down the slope at the side of the road, leaving the others with the two cars. The reason for this, he said, was that if other motorists saw him with his telescope, they would probably stop their cars. He felt this would spoil his chances of seeing anything, presumably because he felt the aliens would steer clear of any crowds.

He wandered about a quarter of a mile into the desert, out of sight of the road, when he suddenly saw a figure waving at him. He thought it was probably a prospector looking for rock samples, and wondered whether the man needed assistance. When he approached, he saw that the "man" was tall, blond, and wearing what looked like a one piece suit, with a broad belt around his waist. It was then that Adamski caught sight of a small UFO some distance away, almost obscured because it was in a hollow. Adamski moved to shake hands, but the tall stranger insisted on a palm to palm greeting. Adamski then tried to communicate orally, but the alien seemed not to understand.

Adamski pointed to the sun, and then indicated numbers one, two and three, meaning the nearest planets to the sun, Mercury, Venus and Earth. Eventually the alien seemed to understand, and stopped Adamski on the number two, which he took to mean that the stranger came from Venus. They then went over to the craft, and Adamski apparently got too close, stumbled, and fell toward it. The alien grabbed him and prevented him from falling under the craft, which would supposedly have resulted in serious injury because of the electromagnetic power that was actively bouncing between the underside of the craft and the ground. Adamski claimed that his arm was momentarily under the craft, and that he often suffered temporary loss of feeling in that arm in later years.

After this first encounter, Adamski claimed to have

been contacted quite frequently, but not necessarily in a physical sense. The aliens were apparently telepathic, and could contact him at any time. It also transpired that the aliens could speak English perfectly, but had chosen instead to open a dialogue through sign language and telepathy.

The aliens' message was straightforward. Atomic tests carried out on Earth were producing dangerous fallout, threatening the other planets in the solar system, which were themselves home to intelligent life. The tests must therefore be stopped. Skeptics have pointed out that there was nothing new here. Indeed, the classic science-fiction film *The Day the Earth Stood Still*, made in 1951, had featured a plot where an extraterrestrial visitor and his robot landed in Washington DC to deliver an identical message.

Another of the aliens' concerns was that we were not yet ready for the advanced technology behind the flying saucers. During his 1963 London lecture Adamski summed this up as follows:

"I don't think we're quite ready to harness that kind of power ourselves, because if we do, as they are afraid of, in that same way that we might put guns on it, and there would be no more peace in space, while they do have peace in space, up until now. And when we decide we're not going to put guns on, I think we're going to be handed that knowledge where we can build such a ship, and finally go to those planets, and see for ourselves."

Adamski was to make further claims, detailing later contacts with extraterrestrials, in his second book, *Inside the Spaceships*, which was published in 1955. As one can deduce from the title, Adamski claimed that he had actually been invited on board some alien spacecraft, at the behest of his extraterrestrial friends. But Adamski was never one for half measures, and not content with having been on board, he had also gone on various jaunts across the solar system. He explained how he had been on a trip around the moon, and seen the far side, which is permanently hidden from us on Earth. He saw forests, mountains, lakes, and even people. The scientific world scoffed at Adamski, explaining how the far side of the moon could not possibly be as he claimed it to be, citing all manner of technical points. My father recalls one television debate when the interviewer put just such objections forward. Adamski seemed bemused by the whole affair, turned to the man and said: "But it is like that; I've seen it." Short of calling Adamski a liar to his face, there was nothing the poor interviewer could do.

In 1959 the Soviet Luna 3 probe brought back the first pictures of the far side of the moon, which showed a bleak and barren terrain. Adamski had an explanation for this: it was all a communist plot. Those sneaky Soviets had obviously doctored the photographs in order to deceive the West!

Adamski's claims became increasingly bizarre, and included accounts of trips to Venus, Jupiter and Saturn. He claimed that the space people traveled around in huge "motherships," several miles in length, and landed in "scout ships" (i.e. flying saucers) that were normally stored within the mothership. Adamski produced a number of photographs purporting to show such craft, along with an

8 mm color film, taken in Silver Spring, Maryland. As with all such material, controversy surrounds its authenticity. Although widely condemned as fakes, the truth is that we simply cannot prove the issue either way.

Adamski is now generally regarded as a fraud, largely on the basis that our own space program seems to have found him out. None of the planets he claimed to have seen at close quarters bears any resemblance to his descriptions, and unless we take the extreme view that virtually all the planetary data gathered by NASA and by the Soviets has been fabricated, we are left with the inescapable conclusion that Adamski was lying. Or are we?

Let us consider an analogy. Suppose you are the first person to visit a strange land, which I shall call Nowhereland. You fall into the company of a stranger, who tells you that she will take you to the capital city, Anytown. While there, you notice that the city is built around the shores of a large lake, and when you return home, you tell everyone this fact. Some time later, many more people visit Nowhereland, and some of them stay in Anytown, and bring back pictures. There is no lake, and you are denounced as a fraud. My point is that, in my scenario, you were not lying about your experience. The stranger had lied to you, and the place to which she took you was not Anytown but somewhere else entirely. Could it be that Adamski genuinely did meet extraterrestrials, but that for whatever reason, they were not being truthful with Adamski?

Now let's have a look at the other famous contactees, and see if we can discern any pattern in their experiences.

George Van Tassell

Van Tassell was born in 1910 and spent his early career in the aviation industry working as a flight instructor and flight safety inspector for companies such as Douglas and Lockheed, before opening Giant Rock Airport in California in 1947. He had also test-flown aircraft for the American millionaire Howard Hughes. Van Tassell claimed that his contacts with entirely benevolent space people included a trip on a flying saucer and a journey back in time, during which he was shown some of Earth's history.

In 1953 he formed an organization called the Ministry of Universal Wisdom, and among his projects was an annual UFO gathering known as the Giant Rock Convention. Perhaps his most bizarre initiative was the construction of a huge machine that he claimed would halt and even reverse the aging process. He maintained that the plans for this machine had been given to him by the space people, but perhaps predictably, it has never worked.

Truman Bethurum

Truman Bethurum published his story of extraterrestrial contact in 1954, in a book entitled *Aboard a Flying Saucer*. His claim was that he had been carrying out road maintenance on a highway running through the Nevada desert when he encountered five aliens, who had arrived in a massive spacecraft. He described the extraterrestrials as being olive skinned, about five feet tall, and wearing uniforms. They spoke perfect English, and looked sufficiently human to move around freely on Earth, which they did from time to time.

Bethurum said that the craft was captained by a female named Aura Rhanes. He described her in various different ways, all of them complimentary: "Queen of women" and "Tops in shapeliness and beauty" being two examples. Some time after the initial encounter, Bethurum saw Aura Rhanes drinking a glass of orange juice at a restaurant. If Bethurum had been making the whole account up, one would expect there to have been further contact here, or even for a story to develop in which the two became lovers. But Bethurum's frank admission, which has the ring of truth to it, is that she ignored him.

Bethurum's aliens came from a planet called Clarion, which was supposedly free of disease, war, or any other perils. Clarion was said to be quite close to Earth, but permanently hidden behind the moon. Modern space exploration has not revealed such a celestial body, but as with Adamski, it is entirely possible that Bethurum *did* have an extraterrestrial encounter with aliens who were not truthful. Why would aliens lie to us about the location of their home planet? One reason might be that when we develop viable space travel, they do not particularly want us coming to visit them.

Daniel Fry

Another book published in 1954 was Daniel Fry's *The White Sands Incident*. Fry was an instrument specialist working on missile-control systems at the White Sands Proving Ground in New Mexico, where the American military tests much of its new hardware. Fry stated that his encounter had been in 1950, although in later years he said this was a mistake, and that the incident had occurred in 1949. One day he saw a huge UFO land in an isolated area near White Sands. He cautiously approached the landed craft, and was reaching out toward it when he heard a voice (speaking in colloquial English) call out to him. But the first words from the aliens were not to be some momentous message from across the void of space, but something a little more practical:

"Better not touch the hull, pal; it's still hot!"

Fry was told that the voice belonged to an entity called A-lan, who was in fact in a much larger mothership, in orbit around Earth. The UFO that Fry could see was apparently a smaller craft used for carrying cargo. A-lan told Fry that they were descended from survivors of a calamitous war between the two legendary "lost continents" of Atlantis and Lemuria, who had migrated out into space after the conflict. This is an inverted version of the so-called "Ancient Astronaut" theory, which holds that extraterrestrial visitors came to Earth long ago, and were worshiped as gods. Variations on the theory—which was popularized by Erich von Däniken—state that these extraterrestrials interbred with primitive humans, or even seeded life here in the first place.

Fry was invited onto the saucer, told about the propulsion system, and then given a return trip to New York, which took only half an hour.

Fry's aliens, like those of most of the contactees, were benevolent, but concerned at some of our antisocial habits. They were particularly worried about the dangers of a nuclear war, and had apparently charged Fry with spreading word of this to the world. This Fry duly did, first through his book, and secondly through forming, in 1956, an organization called Understanding, aimed at bringing

THE SPACE PEOPLE   29

about an enlightened attitude on Earth. Fry gave up his job, and devoted himself to lecturing on his experiences and running his new organization.

Fry was repeatedly asked why, if the aliens wanted to effect change on Earth and deflect us from our nastier habits, they had not made more open contact with the authorities, so that the proof would be there for all to see. He replied that the aliens felt this would adversely affect the balance of power on Earth and undermine the authority of our institutions. The changes would have to be made gradually, through the actions of enlightened individuals. The aliens could not simply come and solve all our problems for us. This would be analogous to a well-meaning child breaking open a bird's egg before it is ready to hatch, to save it the effort of breaking through the shell. In such a scenario the chick will not flourish, because all the time it is in the egg it continues to develop, and even breaking out of the shell serves to strengthen its muscles. It is apparently the same with humans: we have to take responsibility, and help ourselves.

Elizabeth Klarer

Elizabeth Klarer was born in 1910, and her varied career has encompassed such diverse jobs as piano teacher, meteorologist, light aircraft pilot and intelligence officer in the South African Air Force. She claims numerous encounters with benevolent extraterrestrials, mainly between 1954 and 1963. The central part of her story involves an encounter with an attractive male humanoid, with whom she fell in love. She apparently became pregnant, and before the birth she was taken to the alien's home planet, where her baby was born—and supposedly still lives. The planet, like so many of those described by the contactees, is free from war, poverty and disease. Such claims inevitably lead to suggestions that the contactees are constructing escapist fantasies, rather than recounting actual events.

Orfeo Angelucci

Angelucci's encounters were of a much more spiritual and religious nature than those of most other contactees. Angelucci was an aircraft mechanic, and saw his first saucer when it landed in a field near Los Angeles in 1955. Like many of the other contactees, he was given a flight in a flying saucer, and had its propulsion system—the conversion of electromagnetic force—explained to him. The purpose of the contact was a familiar one, and involved the aliens delivering a warning about the dangers of technological advancement when not accompanied by spiritual evolution. The aliens, in other words, were concerned about our nasty toys (especially the military kind).

Subsequently, Angelucci began to meet these aliens in everyday situations, unconnected with any UFO activity. Rendezvous would take place in the most mundane places, such as cafés and bus stations.

Perhaps the most significant contact was with Jesus, who revealed that he was himself an extraterrestrial. Jesus told him that the extraterrestrials were living on Earth, and mixing freely with humans, from whom they were physically indistinguishable. They were here to bring about the New Age. This concept of aliens living among us is very closely related to the more recent idea of a Walk-in, popularized by the American psychic Ruth Montgomery.

Walk-ins are alleged to be intelligent entities who, with permission, swap places with human souls who wish to depart for *elsewhere*. Some Walk-ins are said to be spirits, while some are extraterrestrials who can either swap places with a soul who wishes to depart, or even choose to be born into a human body. Their mission is supposedly to lead us into an enlightened New Age.

As well as his physical journey in the flying saucer, Angelucci said that he had been on a *spiritual* journey to the aliens' planet of origin. There, he met a male alien called Orion, and a beautiful female called Lyre. They told him that in a past life, he had been an extraterrestrial called Neptune. As we shall see, this concept of dual human/alien identity was to resurface nearly forty years later in hypnotic regression sessions carried out by Dr. John Mack, Professor of Psychiatry at Harvard Medical School.

Angelucci relayed an ultimatum from the extraterrestrials: if the human race did not mend its ways, a calamity would occur in 1986—a date that tied in with the next close approach to Earth of Halley's comet. Angelucci's supporters will doubtless be encouraged that we have passed this date without incident; while skeptics will say it simply proves that he was making it up.

Howard Menger

Howard Menger claimed that his contacts with extraterrestrials dated from 1932, in his early childhood, when he and his brother were visited by a beautiful female alien near the family home in New Jersey. He was apparently told that the extraterrestrials would always be with him. Menger was to encounter such beautiful aliens on a regular basis, and on each occasion he would sense great love from them, especially the females, who were fond of wearing fetching one piece jump suits. All the aliens were tall and good looking, although it transpired that even the young-looking ones were several hundred years old.

Depending upon your viewpoint, Menger's encounters are the most bizarre, humorous or ludicrous of all. On one occasion, Menger gave some of the female aliens some bras, only to be told that they did not wear such garments. Menger also claimed that he would regularly cut the aliens' long hair, so that they could move around on Earth without attracting attention. In exchange, he was given a trip to the moon, where he claimed that the atmosphere was similar to Earth's. He said that he had brought back some "lunar potatoes," which were subsequently confiscated by the government! In common with other contactees, Menger was also given a lecture on the propulsion system of the alien spacecraft.

Menger's aliens had apparently visited Earth before, and had helped with the development of a number of ancient civilizations, among them the Aztecs.

In a new twist, Menger started marketing some music that his alien friends had supposedly taught him. *Music From Another Planet* was available through mail order, and doubtless served as a useful source of income, in addition to the royalties from the inevitable book which he wrote about his experiences.

Sid Padrick

Like Orfeo Angelucci, Sid Padrick was to claim contacts that were markedly religious in nature. Padrick's story begins on January 30, 1965 near his California home, when he encounters a landed flying saucer. He heard a voice telling him not to be afraid and that nobody would harm him (echoing the biblical "fear not" message given to the shepherds by the angel of the Lord). He was invited onto the craft, and given a tour. The aliens were essentially human in appearance, albeit with rather pointed chins and noses. Much of what Padrick was told was by now standard contactee fare: the visitors came from a planet that was in our solar system, but permanently hidden behind one of the other planets. Their planet was another of those utopian worlds where there was no crime or disease, and where everybody "lived as one."

But the central part of Padrick's experience, and the feature that sets his case apart from hundreds of others, concerns the invitation that he received while in a part of the craft known as the "consultation room":

"Would you like to pay your respects to the Supreme Deity?"

After ascertaining that the aliens did indeed mean God, Padrick knelt and prayed. He maintains that he genuinely felt the presence of God that night.

Sid Padrick reported his experience to the United States Air Force, and was interviewed at great lengths by personnel from Hamilton Air Force Base, who forwarded their report to Project Blue Book staff at Wright-Patterson Air Force Base. Although they suspected fraud, the case file reveals one particularly intriguing piece of evidence. The Air Force took notes of the conversation that Padrick had supposedly had with the aliens, and at one point he had asked how the saucer had evaded radar detection. The alien replied:

"The hull of our spacecraft absorbs energy, and will not allow a reflection . . ."

Remembering that this was written in 1965, there are remarkable parallels between this description and the radar absorptive paint used today on stealth aircraft!

Billy Meier

Eduard "Billy" Meier is one of the last of the old-style contactees, and his story emerged in 1975. Although stories of benevolent contacts were still being made (and they have never completely disappeared), increasing numbers of abductions were being reported by the time Meier came forward, and the contactees were old news. Despite, or perhaps because of this, Meier's case is one of the most controversial, and has aroused strong passions among believers and skeptics alike.

Meier is a Swiss farmer, whose story centers around his alleged contacts with aliens from the Pleiades and his production of literally hundreds of relatively clear, color photos of UFOs. Many ufologists regard these photos as simply too good to be true, and believe them to be fakes. The skeptics point to numerous inconsistencies in his story, and say that the case is the biggest hoax of all time. But Meier had strong support from ufologist Wendelle Stevens and from a group of militant believers who collectively styled themselves as Genesis III.

Meier's aliens were human in appearance, highly advanced, and apparently took him on trips to the past and the future, as well as introducing him to Jesus. Nowadays

THE SPACE PEOPLE   35

most ufologists regard this case as a hoax, both because of
the increasing bizarreness of Meier's claims, and because
of question marks raised over his photographs by modern
photographic analysis and enhancement techniques.

Analysis of the contactees

The cases just outlined are by no means the only
ones. They are, however, the best known, and taken together, they illustrate the overall pattern of the phenomenon. Some common themes will be readily apparent.

With regard to physical appearance, similarities are
considerable, despite some minor differences. The aliens
are almost always described as humanoid in appearance,
very attractive, often with long, flowing, blond hair. This
type of entity is commonly known as a Nordic.

Communication between the contactees and the beings is often through telepathy, and does not necessarily
require the two parties to be together; after the initial
encounters, many contactees said that they received telepathic messages from the aliens, without having had another face-to-face encounter. This echoes popular stories
of saints and other devoutly religious people hearing
voices. Again, both ancient and modern reports could describe the same phenomenon, perceived in different ways.

The contactees' aliens are all benign, and usually concerned with the danger that Earth's technology poses both
to ourselves and to the wider galactic neighborhood. The
contactees feel that they have a mission to bring this message to the world, and indeed are sometimes specifically
told that this is the reason they were contacted.

A curious point is that the contactees seem almost

exclusively to be American. Contact with aliens can often appear to be a peculiarly American phenomenon. Whether this reflects the true geographical distribution of such experiences is open to debate. On the one hand, if extraterrestrial visitors were genuinely concerned about some of our nastier scientific discoveries, America is where they would probably find the most advanced technology. But, equally, it may simply be that these encounters occur fairly evenly around the world, but are *reported* more in the United States, where the extent and tenacity of the media facilitates the publishing and discussion of such claims.

Many have argued that all the contactees were frauds, cashing in on the emerging public obsession with flying saucers. There were certainly motives for fraud. Each of the cases dealt with in this chapter involves people who emerged from obscurity into the public eye, most of whom wrote books based on their experiences. The twin attractions of fame and fortune are undoubtedly considerations that we must bear in mind.

And there was another more benign reason for fraud: it may be that some of the contactees were genuinely concerned about the potential threat of nuclear war, and felt that the only way that they could ever gain a platform, or stand any chance of bringing about a change, was to project these concerns onto extraterrestrials.

It is interesting to look at how the contactees were viewed at the time. The media were fascinated, and the subject was one that filled many column inches in newspapers and magazines, as well as spawning numerous books. Serious UFO researchers were dismissive, and tried to disassociate themselves from the emerging contactee phenomenon. Dr. J. Allen Hynek, Scientific Consultant to the United States Air Force's official study into UFOs, Project Blue Book, regarded most contactees as "pseudoreligious fanatics" of "low credibility value." Hyneck differentiated between the contactees and people who had experienced what he called a "Close Encounter of the Third Kind," which he defined as a UFO sighting where "occupants" were seen in or around the craft. Hynek regarded this latter phenomenon as interesting, but very rare. In such cases the "occupants" almost invariably eschewed contact. Edward Ruppelt—one of the heads of Project Blue Book—was similarly unimpressed with the contactees. Many ufologists believe it was this that led him to reverse his position in the second edition of his book *The Report On Unidentified Flying Objects*, in which he argued against an extraterrestrial explanation for UFOs. The first edition had left the question open.

It is possible that the contactees may have been the victims of a terrestrial hoaxing and disinformation campaign. Such a campaign might have had the aim of throwing serious UFO researchers off the scent, discouraging scientific study, and generally muddying the waters. We know from documents released in America under the Freedom of Information Act that various military and intelligence agencies were taking an interest in the subject. The US military wanted the secret of the exotic propulsion system (and perhaps weaponry) that would undoubtedly be incorporated into any genuine extraterrestrial craft. They also wanted to make quite sure that nobody else got there first, so they might have had motives for discrediting ufology in general, lest the Soviets thought they were on to something. It is also clear that embryonic UFO groups were regarded as potentially subversive organizations, so that discrediting them had two benefits as far as officialdom was concerned. What better way to undermine the developing science of ufology than to bog it down in humorous anecdotes about lunar potatoes, bras and hairdressing services? Who would ever take such material seriously?

History has not treated the contactees well. Their descriptions of planets—known and unknown—within our solar system have been overtaken by scientific knowledge. They are regarded with suspicion, and often vilified as out-and-out frauds. If any of the cases are genuine, the only theory that then makes sense is the idea that the extraterrestrials were deliberately feeding the contactees false information. But is this so unlikely? Humans lie, so why shouldn't extraterrestrials? The motive for such falsehood might be to avoid us finding out too much about them and their homeworld, for fear we will one day come knocking on their door. Alternatively, it may be a more subtle form of disinformation, designed as a smokescreen for the aliens' *true* motives.

Many have argued that the contactees were a product of their time, reflecting post-war concerns about nuclear war and pollution. They anticipated both the hippie movement and the religious cultists who were to proliferate a decade or so later.

There is another possibility. Could these encounters not be just as real as folkloric encounters covered in the previous chapter? Again, perhaps this is the same, familiar phenomenon, but reinterpreted according to the prevailing culture. The key cultural factors for the contactees were the developing flying saucer mystery, and the depiction of this mystery in classic science fiction films such as *The Day the Earth Stood Still*; so any encounters with non-human intelligences may well have been perceived by them in ways that reflected these factors.

But despite what they have in common, I do not believe we can treat the contactees as a single group, whose claims must be either completely true or completely false. We must never forget that they are a collection of individuals. The truth is seldom as black and white as the die-hards believe, whether they be skeptics or believers.

Generally speaking, the era of the contactees was the fifties. But the sixties brought another change in the evolving saga of human contact with these *other* intelligences. The concept of benevolent space people was about to be challenged by the emergence of some sinister characters with a much more unpleasant agenda.

CHAPTER 3

The Space People Turn Nasty

I f the contactees had simply been making up their claims, then we might have expected more and more such people to come forward, with further tales of the benevolent space people, and increasingly exotic accounts of travel to distant worlds. But this did not happen. What took place was a radical shift in emphasis, with extraterrestrial contact now occurring without the consent of the human participants, and often without much conscious recollection of events. Open contact seemed to have all but stopped, as suddenly as it had started. This in itself seems to me to vindicate at least some of what the contactees had said, and implies that there is a genuine, evolving phenomenon at work here, as opposed to mere human invention. Clearly many of the contactees (and there were many hundreds lesser known than the few whose cases were outlined in the previous chapter) were either lying, or at the very least exaggerating, but I believe that *something* extraordinary did happen to some of them. Perhaps they were the lucky ones, because extraterrestrial contact for others was about to become rather less pleasant.

Setting aside the fairy abductions of folklore, the first modern account of what came to be known as an alien abduction emerged from Brazil in 1965, although the events concerned had supposedly happened in 1957.

Antônio Villas-Boas

Prior to the actual abduction, Villas-Boas had two UFO sightings that are worthy of note in themselves. On the first occasion, on the night of October 5, 1957, the twenty-three-year-old law student and his brother saw a bright beam of light shining down from the sky outside his bedroom window. The second sighting occurred nine days later on October 14, when Villas-Boas was driving a tractor, plowing a field at night. (Although he was studying law, he also had to work hard on his father's farm.) He saw a bright light that appeared to be hovering over the field he was in, so he tried to get closer. The object seemed to dart around rapidly, leading him on a merry chase. Exhausted, Villas-Boas gave up pursuit. He then saw the light send out multiple beams, before disappearing completely. As with the first sighting, his brother had also witnessed the extraordinary light display.

The following night Villas-Boas was alone in the same field when he saw a red light coming directly toward him. On this occasion he was clearly able to make out a structured craft, and this craft—looking vaguely egg-shaped—landed, resting upon three legs. For the first time in the series of sightings Villas-Boas was afraid, and drove away from the landed craft. His tractor stalled, and Villas-Boas was grabbed by four aliens, who rushed out of the craft, seized him, and forcibly took him on board. The aliens were small, with disproportionally large heads—a description that we will be encountering with increasing frequency.

Villas-Boas was undressed against his will and smeared with a strange viscous liquid. The beings then cut him, and took a sample of his blood before leaving him alone in a small, round room. A foul-smelling gas seemed to be coming into this room, and Villas-Boas promptly gave the aliens a souvenir of his visit by vomiting. Some time later a naked female being came in. She looked human, although she had small feet and a wide face, narrowing to a point at the chin. She had piercing blue eyes, long, thin hands and stood about four and a half feet tall. Her hair was blonde, although her pubic hair and the hair around her armpits was red. Villas-Boas found himself becoming aroused, and they had sexual intercourse twice. She took a sperm sample, storing it in a test tube. Before she left, the woman smiled, pointed to her stomach, and pointed upward. The message was clear; a human/alien baby would be born in due course, and raised *elsewhere*. Villas-Boas had the following remarks to make about this part of the experience: "That was what they wanted of me—a good stallion to improve their stock. In the final count that was all it was. I was angry, but then I resolved to pay no importance to it. For anyway, I had spent some agreeable moments [although] some of the grunts that I heard coming from that woman's mouth at certain moments nearly spoilt everything, giving the disagreeable impression that I was with an animal."

After his sexual experience, Villas-Boas suddenly realized that he had a chance to try to obtain some proof of his abduction, and resolved to steal a small device of some sort from the craft. He was discovered but, despite this attempted theft, the aliens gave him a brief tour of the craft, before taking him back outside. The craft then ascended slowly into the air, before moving off at incredibly high speed. The whole encounter had taken a little over four hours.

Over the next few weeks, Villa-Boas developed strange wounds on his arms and legs, which then became scars.

There is no obvious motive for fraud in this case. Villas-Boas did not write a book, and had no intention of gaining anything other than an explanation. Although he reported his experience to Olavo Fontes, Brazilian representative of the Aerial Phenomena Research Organization (a worldwide UFO group formed in 1952) as a result of a newspaper advertisement seeking details of UFO sightings, he was reluctant to give many details of the experience, which had to be coaxed gently from him by Fontes. Villas-Boas never sought or obtained either publicity or money as a result of his encounter, and indeed made no public comment at all until 1978, when he confirmed the

facts of his encounter. Nor did he form or associate himself with any cult or UFO organization, and indeed had no
reason to do so: unlike the contactees, there was no message of greeting or warning to convey. His experience, far
from being one that placed him in a central role as inter-
mediary between two worlds, was embarrassing and de-
grading. Villas-Boas went on to qualify as a lawyer, and
spent the rest of his career in the legal profession.

This case was something entirely new, but was stu-
diously ignored by much of the ufological establishment,
mainly because it did not seem to fit the pattern even of the
contactees, let alone the more routine UFO reports. Despite this, it is clear that the case was indeed a turning
point, not only because it involved the use of force, but
also because of the themes of sex and breeding—themes
that were to come to the fore in later years.

Notwithstanding the Villas-Boas encounter, the
modern abduction mystery really starts with a case that
occurred in 1961, and was made public in 1966. It contains many of the factors present in modern-day cases, and
was to become the yardstick against which all other abductions would be measured. It is worth having an in-
depth look at this case, which is probably the most famous
abduction on record.

Betty and Barney Hill

In September 1961 Betty Hill was working as a social
worker for the State of New Hampshire, specializing in
child welfare, while her husband Barney worked for the
United States Post Office. They had been taking a driving
holiday in Canada, seeing Niagara Falls, and visiting

Montreal. There they heard a weather report that talked about the possibility of a hurricane hitting the coast of New Hampshire. This was worrying news, as the Hills lived in the New Hampshire town of Portsmouth. Abandoning their plans to stay overnight in Montreal, they decided to go straight home, fearful of being cut off by the approaching hurricane. They set out that evening, crossed the Canadian/American border, and arrived in the New Hampshire town of Colebrook at about 10 p.m. that night. At a restaurant, they debated whether to stay overnight in Colebrook, or whether they should drive through the night, and stop for a rest if they became tired. The distance from Colebrook to Portsmouth is about one hundred and seventy miles down US highway 3, and so, after a brief snack, the couple decided to press on. They left the restaurant at 10:05 p.m., and Barney, who was a careful planner of routes, estimated that they should arrive back home at 2:30 a.m., or 3 a.m. at the very latest. In the event, they did not arrive back home until 5 a.m., a fact that was to have the most profound significance for the entire study of the alien abduction phenomenon.

Just to the south of Lancaster, Betty noticed a bright object in the sky, immediately to the left of the full moon, and not far from Jupiter. The object seemed to be moving, getting bigger and brighter, although it was difficult to tell whether the movement might not have been due to the motion of the car. Betty told Barney, who initially thought that the object might be a satellite. Shortly thereafter, the Hills' little dachshund, Delsey, began to get restless (there are numerous reports of animals being spooked by UFO activity), so they decided to pull over to let her go for a walk. Once out of the car, they were able to get a better look at the object, and were now certain that it was indeed moving.

They got back into the car, and continued their journey, all the while wondering what the object could be. Betty was insistent that it was not a satellite, and Barney eventually offered the view that it might be a commercial aircraft. The object seemed to be following them, and they made periodic stops in an attempt to identify it. By now it was clear that the object was neither a star, satellite or aircraft. It seemed to hover, motionless, and then move with bursts of speed. It had multicolored lights, and was flashing out beams of light. Through the binoculars they could make out a fuselage, but no wings. There was no engine noise at all. They drove past a resort area at Indian Head still speculating about the unidentified craft. Barney was a nervous man, and was beginning to get tense and worried. The roads had been completely deserted for many miles now, and Delsey was whining and hiding under the seat. At this point the object suddenly came much closer, and from a distance of no more than two hundred feet, Betty saw that the object was a huge, structured craft with a row of windows.

At Betty's insistence they stopped the car in the middle of the road, and Barney took the binoculars. He began to wander off the road and into a field on the left hand side of the car. Betty had been concerned that the car was directly in the middle of the road, and had remained sitting in the vehicle, watching for other cars. When she noticed that Barney had wandered off, she shouted at him to come back, but he didn't hear her. He was looking at the object through the binoculars, and this time he saw about half a dozen figures through the row of windows, all dressed in what appeared to be shiny black uniforms. The craft was getting nearer, and some sort of extension seemed to emerge from the underside; clearly it was going to land. All the figures except one then dashed to a control panel of some kind, leaving only one figure at the window, staring intently at the terrified Barney.

Barney seemed mesmerized by the eyes of the being, but through a supreme effort managed to look away. As if a spell had been broken he ran, screaming, toward the car, having been gripped with a fear that he was about to be captured. He leaped in, and drove off at high speed. He asked Betty to check whether the object was in sight, but it had disappeared. Shortly thereafter, the couple heard a strange series of electronic beeps, and after this, their memories became hazy, until they heard a second series of beeps, which restored their awareness. They were now about thirty-five miles south of Indian Head, with little recollection of the preceding events. The feeling had been like regaining consciousness, but they had been driving the car, so there was so way in which sleep could explain their peculiar, shared feeling. They continued their journey, and when they got back to Portsmouth it was 5 a.m. Their journey had taken two hours longer than it should have done, although this only emerged much later, while their case was being investigated.

After a light snack, Betty and Barney went to bed, and slept until about 3 p.m. in the afternoon. The previous night's events were fading from their memories, and Barney was thankful. He pushed the sighting of figures aboard the craft out of his mind, and resolved to forget the whole thing. He didn't want to remember. Betty did want to remember, and decided to share what she could recall of events with her sister, Janet. Janet had a neighbor who was a physicist, and sought his advice. She then advised Betty to check the car with a compass, to see whether there was any unusual radiation. (It is uncertain exactly what was said, but clearly the message became garbled somewhere, because one cannot check for radiation with a compass.) Betty went to the car, and noticed about a dozen shiny, round spots on the trunk. Around these spots, the compass began to spin wildly. Elsewhere the reaction was normal. Betty ran to tell Barney, but in his eagerness to forget the whole affair, he showed little interest—much to Betty's annoyance.

At the suggestion of Janet—who had spoken to a friend who was a former Chief of Police—they reported their sighting to nearby Pease Air Force Base, although they left out details of the shiny spots on the car and the figures Barney had seen. It has since emerged that an anomalous radar contact *was* detected in the area of their sighting on the night in question, at a time (2:14 a.m.) that could well tie in with the Hills' encounter. This report was duly forwarded to the staff of Project Blue Book at Wright-Patterson Air Force Base in Ohio.

Betty began to wonder more and more about the mystery of UFOs, and obtained a book on the subject from the library. The book was *The Flying Saucer Conspiracy* by Donald Keyhoe, a former Marine Corps Major who was now Director of the National Investigations Committee on Aerial Phenomena (NICAP), a leading UFO research group of the day. Some days later, Betty began to have a disturbing series of dreams in which she was taken onto a spacecraft and medically examined. It was only later that it was suggested that these dreams might reflect events that had actually happened.

Prompted into action by her dreams, she wrote to Keyhoe, and before long, the whole incident was being investigated by NICAP. Walter Webb, a Scientific Adviser to NICAP, interviewed the Hills at length and was impressed by the way in which they consistently downplayed events. Further investigators were brought in, and during another interview with the Hills the idea of using hypnosis to determine what had happened in the "missing" two hours was floated. Interestingly, the idea was put forward not by the NICAP investigators, but by James MacDonald, a family friend of the Hills who had recently retired from the United States Air Force.

By the summer of 1962 Barney began to suffer from exhaustion, high blood pressure and ulcers, and his doctor recommended that he see a psychiatrist, Dr. Duncan Stephens. At first, Barney hardly ever mentioned the UFO sighting, but eventually, in a bid to get to the bottom of his problems, he was a little more forthcoming about their encounter. He then mentioned to Dr. Stephens the idea of hypnosis, having again had the procedure recommended to him, this time by Captain Ben Swett from nearby Pease Air Force Base, who had met the Hills at their church discussion group. Dr. Stephens then recommended the services of a distinguished Boston psychiatrist, Dr. Benjamin Simon.

The first consultation with Dr. Simon took place on

December 14, 1963. Betty accompanied Barney, and since they had both had the experience, and in view of Betty's dreams, Dr. Simon felt that both of the Hills needed treatment. The first hypnotic session was held on January 4, 1964, and the final one on June 27, 1964. Between these two dates the hypnotic regressions carried out on Betty and Barney were to produce the most extraordinary account of what happened to them during the period of amnesia that occurred in between the two sets of electronic beeps that they had heard.

Dr. Simon carried out regression sessions with both Betty and Barney, but conducted these separately, to try to avoid one account becoming contaminated with details from the other. Again, to avoid the danger of contamination, he tried to ensure that the material recalled under hypnosis was not remembered afterward, until toward the end of the sessions, when he played tapes of the earlier regressions to the couple, to see if they could come to terms with the events.

The story that emerged under hypnosis was that after driving away from the area where he had sighted figures in the craft, Barney then inexplicably turned off the main road a few miles farther south. He headed toward what he thought was the setting moon, only to find the road blocked by a group of figures. One had a device of some sort in his hands, and at that moment the car engine cut out. Barney thought they were about to be robbed, while Betty thought that the men's car had broken down. Betty suddenly sensed that something was wrong, and tried to run, but was grabbed by some of the men. It was then that she saw that they were not human. Barney described them as humanoid in appearance, with very high foreheads, and distinctive, slanted eyes. "I didn't think of anything. I didn't think of the man in the sky in the machine that I saw, I just saw these eyes, and I closed mine." He was incapacitated in some way, and seemed to have his eyes shut for the rest of the encounter. Betty struggled to stay awake and alert, fighting against some unknown compulsion. They were then taken onto the landed craft, placed in separate rooms, and given medical examinations. Betty recalled that she had a long needle inserted through her navel, and the procedure was described to her as being a pregnancy test. She was told that they would come to no harm, and that all the beings wanted to do was to test the differences between the two species.

Perhaps the most remarkable information to emerge from the hypnosis sessions was Betty's account of what came to be known as the Star Map. The aliens had apparently shown Betty a map of trade routes and expeditions, and Betty had asked where their home planet was. The leader replied by asking Betty where Earth was, and when she said she didn't know, the leader being retorted that there was no point in his telling her where *they* came from if she didn't know where *she* was on the map. Under hypnosis, Betty drew the map that she had seen, and some years later an amateur astronomer named Marjorie Fish found a potential match for Betty's Star Map when studying data from a newly published star catalogue. The stars at what appeared to be the start of the map were Zeta Reticuli, a binary star system some thirty-seven light years from Earth. In stellar terms these twin stars are very close neighbors to our solar system. As we will see, the idea that intelligent aliens are coming to Earth from Zeta Reticuli is one that has cropped up in many subsequent cases.

Betty asked for proof of her encounter, and was given a book from the craft. Later, however, the leader being took it back, and said that it had been decided that they would not remember the experience. Betty became very angry, and promised that whatever happened, she would remember. Subsequently, Betty and Barney were escorted back to their car, and continued their journey, after having watched the craft depart.

Dr. Simon had never come across anything like this in his life, and was fearful that the case would ruin his reputation and totally discredit hypnosis as a tool for mental health professionals. He was quite certain that the Hills were telling the truth, although he had great difficulty in accepting that the events recalled under hypnosis had happened in a physical sense.

The story of Betty and Barney's experience emerged—much to their distress—in a series of unauthorized articles in a Boston newspaper. As a result, the Hills decided to tell their own story, to put the record straight. They collaborated with the author John G. Fuller, whose book, *The Interrupted Journey,* was published in 1966. Dr. Simon made the tapes of the regression sessions available, but only because this was what his patients wanted, and only on the basis that he was given editorial control over the medical statements, to ensure their factual accuracy.

Betty Hill took a lie-detector test once, on a national television program. She was asked whether she had received the information about the Star Map on a UFO, and whether any of the material in her account was hoaxed. She scored a very high rating for truthfulness. She and Barney were, in any case, solid, respectable citizens, active in both the local church and the civil rights movement.

I am in frequent demand on the lecture circuit. I am happy to accept such invitations, presenting as they do an opportunity to correct some of the misconceptions that have arisen about the position of the British Government in relation to these matters. These lectures also give me the opportunity to supplement my earlier official research and to meet the key experts and witnesses. In September 1995 I flew to New Hampshire, where I was due to give a lecture at the New Hampshire Mutual UFO Network Fifth Annual UFO Conference. My hosts, Pete and Fran Geremia, introduced me to Betty Hill, and I had a chance to talk to her at first hand about her experiences.

Betty is a charming and very down-to-earth lady, and I had absolutely no doubt whatsoever that she was being entirely truthful. She had come to terms with the events of September 19 and 20, 1961, despite the undoubted trauma that she had experienced at the time. She seemed almost amused, looking back at some of what had happened, and recalled getting into what she called "a fist fight" with her captors, as they tried to get her on to the craft. She describes the beings not as aliens, but as "astronauts," and in the final analysis is glad that the experience happened. There had been, she explained, much speculation about UFOs, and the possibility of life existing elsewhere in the universe. But, as she said: "We were the *only* people in the world who knew for sure."

If Betty and Barney *had* been the very first people to undergo what we now suspect is a fairly standard procedure, it seemed to me that there must be some special factor about them that had set them apart from others, and made them particularly attractive to any intelligences studying human behavior. Betty is white, and Barney was black, and while there is nothing remarkable about such marriages nowadays, they were unusual at the time, especially in conservative New Hampshire. I wondered out loud whether two people who had defied the racism of the day, and were actively involved in fighting against prejudice might have seemed more enlightened and worthy examples of all that is best about the human race. Betty smiled, shrugged, and simply said: "Maybe. Who knows."

Betty's positive attitude is perhaps best illustrated by a comment at the beginning of her own book, *A Common Sense Approach to UFOs*. As well as being dedicated to the memory of Barney (who died in 1969), Dr. Benjamin Simon and John G. Fuller, the book carries a further inscription: "Dedicated to UFOs with Love."

Finally, it is perhaps worth pointing out that New Hampshire seems to have more than its fair share of UFO reports, and has done for many years. Indeed, in 1993 one of the local papers ran a front page story under the following headline: IF YOU SEE A UFO OVER THE PISCATAQUA RIVER, DO NOT CALL 911. IT IS NOT AN EMERGENCY.

Maybe sometimes it is.

The case of Betty and Barney Hill was a watershed for ufologists. It seemed to mark the beginning of something new, and something rather more disturbing than the benign encounters of the contactees. The Hill case is important for a number of reasons, not least that it was the first to involve the use of hypnosis to unlock memories that might have been forgotten by the conscious mind. Much controversy surrounds this technique, and it is appropriate to address this issue here, so that readers can bear this in mind with regard to other cases where some or all of the information emerges only after the witness has been hypnotically regressed.

Regression Hypnosis

First, I must dispel a few myths about hypnosis. Hypnosis is not a magic way of unlocking hidden memories. It is simply an altered state of consciousness where people are more relaxed, less likely to become distracted, and consequently become more focused on a particular topic drawn to their attention by the hypnotist. I have been hypnotized on one occasion, and although relaxed, I was perfectly conscious throughout, aware of my surroundings, and not under any sort of magic spell, as some suppose.

There are several problems with regression hypnosis in relation to abduction research. First, it is entirely possible for somebody to lie while under hypnosis. Equally, it is possible for people to make mistakes about what they recollect, just as we make mistakes when trying to recall events under normal conditions. But of potentially greater significance is the fact that the subject's suggestibility makes it possible for the hypnotist to lead the witness, albeit unintentionally. This happens because the will of the hypnotic subject becomes suppressed, while the hypnotist suddenly becomes a very central figure in the subject's awareness. Subjects may feel under pressure to come up with information—*any* information—in order not to have wasted the hypnotist's time. This information can be brought forward from anywhere in the mind. It may be genuine, but it could be drawn from dreams, fantasies, or even from books or films that the subject has encountered. There may well be no deliberate deception involved, just what is known as *confabulation*, where a subject becomes talkative, in an attempt to fill any awkward silences.

There can also be a subconscious desire on the part of the subject to please the hypnotist, coming out with information that he or she feels is wanted. Thus, if a potential abductee is hypnotized by somebody with a track record of dealing with such cases, the subject may subconsciously be keen to provide material that fits the pattern. The problem nowadays is that so many researchers into the alien abduction mystery have learned hypnosis, and use it in their investigations, without really being aware of its limitations.

There are other problems that arise from a situation in which hypnotherapy is commonplace and knowledge of abductions is spreading fast. I was horrified to hear of two instances where individuals sought the services of a hypnotherapist, wishing to explore the possibility that they had been the victims of childhood sexual abuse. Before any sessions had started, the hypnotherapist volunteered the suggestion that the subjects might have been abducted by aliens. Although, as we shall see, there is a separate debate about the potential relationship between alien abductions and childhood sexual abuse, it is grossly unprofessional for any hypnotherapist to volunteer such a suggestion out of the blue, before any therapy, counseling or regression has taken place. I am pleased to say that in the two cases where I am aware this happened, the subjects walked out in disgust.

If hypnosis is used at all, I believe it is most effective in the hands of qualified mental health professionals with no special knowledge of or opinions on abductions. Otherwise, it is too easy to lead the witness with questions such as: "Can you see any figures in the room with you?"

A professional hypnotherapist should use a minimum of questions, and should try to encourage straight narrative. When a prompting question is necessary, it should be a neutral one, such as: "What can you see around you?"

To be fair, some ufologists will deliberately throw in a leading question to see if a witness can be led, in order to weed out fraudulent or confabulated accounts.

While I am happy to evaluate material that has arisen from regression hypnosis, I have never used this technique myself while investigating cases of abduction. It is my preference to use data that is consciously recalled, and there *are* plenty of such cases around.

When studying alien abductions, we cannot avoid the controversy that surrounds the validity of regression hypnosis, simply because it is used in so many cases. We will see how this situation arose in the next chapter, but for the moment it is enough to counsel caution. Do not treat data that has emerged from hypnosis as gospel, but do not write it off either. Treat it with the same healthy skepticism that you should apply to accounts that are consciously recalled.

The case of Betty and Barney Hill set the pattern for the modern abduction phenomenon. Now let us have a look at some other early cases.

Maureen Puddy

The skeptics love this case, because they maintain it
proves that abductions do not happen in the physical universe at all, but only in the mind of the witness. I do not believe this case shows anything of the sort, and even if it
did, it still would not undermine the entire phenomenon.
One case alone can neither prove the existence of alien abductions, nor invalidate such claims.

Mrs. Puddy's experiences started on July 5, 1972 with
the sighting of a huge UFO that hovered over her car on a
road near Melbourne in Australia. It was shaped like two
saucers, and she heard a low humming sound coming
from the object, which glowed bright blue, and terrified
her. On July 25 she encountered a similar object in almost
exactly the same place, claiming that her car engine cut
out when she tried to drive off. On this occasion she
started to hear telepathic messages, one of which was the
rather clichéd: "We mean you no harm."

In February 1973 Mrs. Puddy received a telepathic
message telling her to go to the stretch of road where her
previous encounters had taken place. She telephoned two
UFO researchers—Paul Norman and Judith Magee—
whom she had contacted after her previous experience,
and they rendezvoused at the site. Mrs. Puddy was agi-
tated, and explained that on her way to the site an alien
wearing a strange garment that appeared to be made of
gold foil had materialized in her car. Later, while in the
company of Norman and Magee, she shouted out that she
could see the alien, and there then followed an extraordinary narrative in which she described how she had been

kidnapped, and was now inside the UFO. At all times during this narrative she was sitting in the seat of her car, in full sight of the UFO investigators, and at one stage she seemed to be in a trance-like state.

Fraud? A psychological delusion? Or a real event that took place in a different reality, which Maureen Puddy could somehow glimpse, while Norman and Magee could not?

Charles Hickson and Calvin Parker

Hickson and Parker were shipyard workers from Mississippi, who claimed to have been abducted by aliens while they were fishing from a disused pier on the Pascagoula River on October 11, 1973. After hearing a strange buzzing sound the two men were amazed to see a large, egg-shaped craft moving slowly toward them, emanating a blue/white glow. The craft landed, and three strange entities floated out and moved toward the two men. The creatures were about five feet tall, had wrinkled gray skin, and strange jutting protusions in place of ears and noses. There was a very small slit that may have been a mouth. At this point, the nineteen-year-old Parker fainted. The two were taken into the craft.

Hickson was given some sort of medical examination that involved him being scanned by a floating device, which resembled a large eye, but after about twenty minutes, both men were unceremoniously dumped back on the riverbank prior to the UFO's departure.

The two men's story impressed the local sheriff, who at one time left the pair alone in his office, which contained

a hidden microphone. The sheriff reasoned that if the two
had been carrying out a hoax, they would probably have
used the opportunity to go over their stories again, or
simply have a laugh at everyone's expense. This did not
happen. Instead the following exchange took place:

> *Hickson: I never seen nothing like that before in my
> life. You can't make people believe.*
> *Parker: I don't want to keep sitting here. I want to see
> a doctor.*
> *Hickson: They better wake up and start believing.
> They better start believing.*
> *Parker: You see how that damn door came right up?*
> *Hickson: I don't know how it opened son, I don't
> know.*
> *Parker: It just laid up, and just like that those sons of
> bitches, just like that they come out.*

This hardly strikes me as the conversation of two people
involved in a hoax.

The two men were interviewed at nearby Keesler Air
Force Base, and subsequently had their case investigated
by Dr. J. Allen Hynek, who had formed his own UFO
group when the United States Air Force study, Project
Blue Book, was terminated. Regression hypnosis was attempted on Hickson, but he became too distressed to continue the session. (Given the doubts that surround the use
of regression hypnosis, it is important to stress that all the
details mentioned here were consciously recalled.) Parker
was so traumatized by what he had experienced that he
subsequently had a nervous breakdown.

Travis Walton

Few cases illustrate the controversy that can surround allegations of alien abduction as clearly as the case of Travis Walton. Walton was one of a group of seven men working on thinning out an area of the Sitgreaves National Forest in Arizona. On November 5, 1975 at around 6 p.m. the men had just packed up for the day, and were heading back home in the truck belonging to the group's foreman, Mike Rogers. Shortly after they drove off they noticed a strange glow in the trees. As they rounded the corner they saw a disc-shaped UFO hovering over a nearby clearing. They screeched to a halt and Walton leaped out and ran toward the object, which then started spinning and emitting an electronic beeping sound. Suddenly, Walton was struck by a beam of light from the craft, and was sent flying backward. Rather ungallantly, his companions drove off rapidly, in a state of panic, before deciding to turn around and come to Walton's assistance. As they drove back toward the area where he had been hit, they saw a light flashing upward into the sky. There was no sign of Walton. His companions reported him missing to the local sheriff, and a detailed search was conducted over the next two days, without success.

On November 11, Walton called his sister's house from a town twelve miles from the site of his disappearance, and asked someone to come and fetch him. He was found on the floor of the telephone kiosk he had called from, and seemed to be in a highly confused state. When he had recovered sufficiently to tell of his ordeal, he related a story of how, after having been struck by the beam of light, he had been unconscious, but had awoken to find himself in a hot, damp room with a low ceiling. He saw three short aliens, with large heads, large eyes and very small ears, noses and mouths. He became hysterical, tried to attack the entities, and then ran out of the room. He then saw a human, wearing a helmet of some sort, who led him out of the craft, and into a large area rather like a hangar, where a number of smaller discs were parked. Walton followed the man into one of these disc-shaped craft, where he saw three more humans. At this point he had a mask put over his face, which again rendered him unconscious. When he came round, he was lying on a road, and could see the UFO rising up into the air.

A furious debate raged about these events. Skeptical UFO researchers such as electrical engineer Philip J. Klass have argued that the case is a hoax. Klass points out that Rogers was behind with the forest clearance work, and in danger of defaulting on his contract, which would have resulted in a financial penalty. As it happened, this penalty clause was never invoked, on the basis that an "Act of God" had occurred: in other words, the crew were too frightened to return to the site. This has been seen by some as a motive for a hoax, as has the $5,000 that the crew received for their story from the *National Enquirer*.

Controversy surrounds two lie-detector tests that were administered to Walton, one of which he passed, and one of which he failed. His colleagues were also given a polygraph test while Walton was still missing (there was some suspicion that the other members of the crew had killed Walton, buried the body, and invented the UFO story), which five of the six men passed. Doubts are invariably cast upon a case where a test has been taken and failed, but the use of lie-detectors is not an exact science, as many suppose, and indeed the British legal system places no reliance whatsoever upon polygraph results. The difficulty is that the test measures stress, and if Walton had been abducted by aliens, and was recalling the trauma, it is *this* that could have produced the stress, and the consequent suspicion of fabrication. Polygraph tests—like regression hypnosis—do not *necessarily* separate truth from falsehood.

Over twenty years after these events, all seven men are sticking to their stories.

Betty Andreasson

Although Betty Andreasson did not report her abduction experience until 1974, the event itself took place on January 25, 1967. The encounter had apparently started with a power failure (interestingly, there are many reports that seem to link UFOs with massive blackouts, the most famous of which occurred in the northeast of America on November 9, 1965). When the lights in Andreasson's home came back on, she found that the rest of her family had been immobilized, although she was able to move quite freely. All of a sudden, four entities entered the house, walking through the solid front door. They were about three and a half feet tall, with large, slanted, almond-shaped eyes, and tiny noses and mouths. They were wearing skintight uniforms, adorned with an emblem rather like a bird. The leader introduced himself as Quazgaa, and Andreasson was then escorted out of the house and into an oval-shaped UFO, where she was given a painful medical examination. At one stage she was enclosed in a capsule of some sort, which was filled with a strange fluid. Subsequently, she was somehow transported to a place where strange creatures were climbing over buildings; saw a gigantic bird, and heard a voice that told her she had a mission in life that would be revealed to her at a later date.

Much of this information emerged only after regression hypnosis, but parts of the story were confirmed by Andreasson's father and eldest daughter, who had apparently been conscious for at least part of the time that the beings had been in her house.

A great deal of the imagery that Andreasson recalled was religious in nature, perhaps reflecting her strong Christian beliefs. At one point, Andreasson gave Quazgaa a Bible in exchange for an alien book, which was subsequently mislaid.

Analysis of the abductees

These cases are interesting in themselves, but far more interesting to me is the issue of *why* things had changed. Why did we now have abductees instead of contactees? Had the extraterrestrials simply altered the agenda, or moved on to the next part of some great plan, or does the answer lie more within ourselves and the way in which we perceive them? It seems to me that these entities, whatever they are, might in some way mirror the state of society itself. In other words, we see something of ourselves in them. The phenomenon may be entirely real, but like a blank canvas on which we paint our own picture, or like a Rorschach test where *we* interpret the inkblots.

As society became more complex and impersonal, and as we began to lose our sense of community, did the alien phenomenon change accordingly? Our lives are increasingly dominated by computers, fax machines, mobile phones and other such innovations. There has been a loss of control, and an abrogation of personal responsibility, and this is what the modern alien abduction scenario may reflect.

Cases such as those in this chapter show how, in the sixties and seventies, the alien abduction phenomenon first came to the public eye. A pattern was beginning to emerge, but the phenomenon itself was still regarded as one of those quaint little mysteries, like the Bermuda Triangle, or the Loch Ness Monster. There was no indication of the impending storm. Few people could ever have suspected that reports of alien abductions would soon reach epidemic proportions, catapulting the subject out of the domain of the UFO magazines, and into the mainstream.

CHAPTER 4

Big News and Big Business

By the mid-seventies, the concept of alien abduction was gaining recognition, at least in America. Cases such as those in the previous chapter had served to bring the subject substantial media attention, not least with the *National Enquirer*. This tabloid magazine was offering a cash prize of $100,000 for definitive proof of extraterrestrial life, and in 1976 they raised the prize to $1,000,000. They handed out smaller cash rewards of between $5,000 and $10,000 each year for the best UFO story sent to them. Skeptics argue that this provided an obvious motive for hoaxing, and point out that both Travis Walton and

Betty Andreasson contacted the magazine about their experiences.

The various UFO groups had been fairly cautious about abductions, and generally wary of associating themselves with any of the more exotic reports. Some ufologists feared that any scientific credibility that they had gradually earned might be destroyed in an instant if they or their colleagues endorsed abduction reports. And yet times were changing. In 1976 Dr. J. Allen Hynek, the ex–Blue Book scientist who had formed the Center for UFO Studies, appeared on a talk show with Travis Walton, and while he didn't altogether support Walton's claim, he nevertheless acknowledged that Walton's account fitted the gradually emerging pattern.

In 1978, ufologist David Webb wrote an article for the Mutual UFO Network's *UFO Journal* in which he estimated that there had been approximately 150 reports of alien abductions to date. Around two-thirds of these reports had come in during the preceding two years, so it certainly looks as if—for whatever reason—the mid-seventies was the point at which the phenomenon began to make the quantum leap.

One explanation for the increasing number of reports was the number of ufologists now using regression hypnosis as an investigative tool. Dr. Leo Sprinkle, a psychologist, based at the time at the University of Wyoming, was one of the first ufologists to make extensive use of regression hypnosis. Sprinkle, who was a consultant to the Aerial Phenomena Research Organization (APRO), started using this technique in 1967, and by the end of the seventies he had regressed around fifty witnesses to UFO sightings. Over half of these cases produced information that might indicate that an abduction had taken place.

The other ufologist making extensive use of hypnosis in the mid to late seventies was APRO's Director of Investigations, Dr. James Harder. Both Harder and Sprinkle based their approach on that of Dr. Benjamin Simon, who had regressed Betty and Barney Hill. In other words, they looked for periods of missing time, and focused their attention on these, seeking repressed (or perhaps suppressed) memories of an abduction. Sprinkle was a trained psychologist and counselor, so it was generally accepted that he knew how to use hypnosis in a professional manner. Harder, on the other hand, was an engineer, and skeptics who were already suspicious of regression hypnosis were doubly concerned when the technique was used by those who were not qualified mental health professionals. Over the next few years there was to be a veritable explosion in the use of regression hypnosis, most of it carried out by ufologists who had a bare minimum of training before undertaking their own, unsupervised regressions. The skeptics regard this as invalidating *all* the data recovered in this way. I disagree. As long as we remember to treat material that emerges under hypnotic regression as *no more or no less valid than consciously recalled material*, there should be no difficulty. Both methods can uncover the truth, and both can produce untruths.

Every now and then an individual comes forward and revolutionizes a particular field of knowledge. In the field of alien abductions, Budd Hopkins is such an individual and his contribution has probably been the single most important factor in bringing the subject to a wider audience.

Budd Hopkins is a successful New York–based modern artist whose interest in the UFO phenomenon goes back to 1964 when he, his first wife and a friend had a UFO sighting. Over the next few years Budd took a casual interest in the subject, reading a handful of books, and collecting some newspaper clippings. His active involvement began in the summer of 1975 when he investigated a wave of UFO sightings that had occurred near his summer house in Cape Cod. A few months later, in November, Budd stumbled on what may have been an abduction case.

The case came his way quite by chance, when George O'Barski, the owner of a liquor store across the road from Budd's New York home, mentioned an extraordinary experience he had had in January 1975, when he witnessed a UFO in the city's North Hudson Park. When the craft was a few feet from the ground, a number of short figures in one-piece suits and helmets came down a ladder, took some soil samples, and then departed. O'Barski was seventy-two years old, and a highly respectable, down-to-earth character.

Budd examined the case with the Mutual UFO Network's New York State Director, Ted Bloecher, and another investigator, Gerry Stoehrer. Over the next few months they turned up a number of other witnesses to the event, together with the fact that there was some time unaccounted for in O'Barski's account: his journey home on the night of his encounter had taken perhaps as much as half an hour longer than it should have done. Budd was vaguely familiar with this so-called missing time phenomenon from the Betty and Barney Hill case, and wondered whether O'Barski might consider hypnosis to uncover any lost memories from his encounter. O'Barski declined, but Budd was hooked, and so began a long voyage of discovery.

The O'Barski case (and Budd's involvement) was featured extensively in the media, and this prompted people to come to Budd with details of their own encounters. By 1977 Budd had started working with a psychiatrist, Dr. Robert Naiman, who used hypnosis to probe the periods of missing time from some of the UFO cases about which Budd had sought his advice. Then in 1978 came the quantum leap.

Steven Kilburn (a pseudonym) had no recollection of being abducted. He had not even seen a UFO. But he had a vague feeling of unease about a certain stretch of road that he used to drive down, and wondered whether hypnosis might uncover the reason. Under hypnosis, an abduction account emerged: his car was somehow stopped, he was escorted onto a craft, and given some sort of medical examination. The description of the aliens was familiar: about four feet tall, with egg-shaped heads, large, black eyes, and very small noses and mouths. Their skin was whitish, with a hint of gray. But here was the key point: if an abduction experience could happen without the witness even recalling a UFO sighting, then who could say how many other such cases there might be? Ufologists had always regarded close encounters as a small subset of the overall UFO phenomenon, with abductions being rarer still. But suppose this was not the case. What if abductions were not rare at all, but very, very common?

Budd was beginning to notice other trends. One recurring theme was that abductions hardly ever seemed to be once-only incidents. There was often a series of events, on a regular basis, with the first experience having taken place during early childhood. Another common denominator was the high instance of anomalous scars on the bodies of many abductees, who had no conscious recollection of how the original injury had been sustained. Some of these scars were short and straight, while some had the appearance of scoop marks, as if a tissue sample had been taken. This was certainly consistent with the numerous reports of a medical procedure being the central part of many abduction experiences.

Budd continued to investigate cases, with hypnosis being carried out by Dr. Aphrodite Clamar, a psychotherapist colleague of Dr. Naiman's. Budd's first book, *Missing Time,* was published in 1981, and ended with an invitation to anyone who felt they might have been abducted to write to him, care of his publishers.

This invitation opened the floodgates, and Budd received hundreds of letters from all over the world. In 1983 he was contacted by Kathie Davis. This pseudonym is the name by which she is better known to ufologists, although she has now allowed her real name—Debbie Tomey—to be used. Budd began an investigation into what he was later to refer to as the most important abduction case yet to have emerged.

Kathie Davis's experiences, and those of other members of her family, were recounted in Budd's second book, *Intruders,* which was published in 1987. It told of a series of encounters, beginning at around the age of six, and having as their apparent purpose the production of alien/human hybrids. During one abduction, Kathie recalled being impregnated by some form of artificial insemination. Subsequently, she recalled an abduction where the fetus was removed from her womb. There followed a bizarre account of how, during another abduction several years later, she was shown a number of strange hybrid creatures whom she was told were hers, and encouraged to bond with them. Many of Kathie's abduction experiences revolved around procedures that seemed to be medical—and specifically gynecological—in their focus. Budd was by now fairly sure that the explanation for the entire abduction phenomenon was that extraterrestrials were carrying out a long-term genetic breeding program, using humans to provide the material, and perhaps combining it with their own genetic stock to produce hybrids.

Although Budd pursues the truth about abductions with vigor, he never loses sight of the fact that the experiences he investigates can be deeply traumatic. Throughout his work with abductees he is mindful of this, and believes that the aim of each investigation must include helping the individual to come to terms with their experiences. I met Budd when we were speaking at a conference together in 1995, and witnessed his warmth and compassion at first hand. He struck me as being in the same mold as a kindly police detective who has spent years trying to solve a particularly complex case on which most of his colleagues have given up. He seemed driven by the need to piece together the evidence to some great crime. But these were, perhaps, crimes against humanity, and Budd has never once lost sight of the faces of the victims. He felt for them; felt their anguish, and despaired at the injustice of their pain.

Budd's concern and compassion led him to set up an organization known as the Intruders Foundation, aimed at creating a network of sympathetic researchers and therapists; its twin purposes were to help the traumatized abductees and to investigate the phenomenon. As ever, the former took precedence over the latter.

Another influential researcher first came on to the abduction scene in 1986. David Jacobs was an Associate Professor of History at Temple University, and his doctoral dissertation had been on the history of the UFO mystery. His interest dated back to the sixties, but it was his 1982 meeting with Budd Hopkins that encouraged him to get directly involved in abduction research. By 1985 Budd had begun to use hypnosis himself, instead of referring his cases to a psychotherapist. Jacobs had also learned the technique, and in 1986 carried out his first regression. Most of Jacobs' investigations employed regression hypnosis, but crucially, cases where the witness also had conscious recall produced information that was the *same* as that which emerged under hypnosis. Jacobs' findings mirrored Budd's, and seemed to indicate that the focus of the abduction experience involved physical—and sometimes mental—examinations, along with other medical procedures ranging from impregnation to the insertion or removal of implants (tiny devices that have allegedly been found inside abductees' bodies, and may be tracking devices of some sort). Like Budd, Jacobs felt that the aim was probably to create a hybrid race—on the aliens' terms, and not for any human benefit. This and other theories about the *purpose* of alien abductions will be explored in chapter sixteen.

There was another alien abduction book published in 1987, and in terms of sales it overshadowed Budd Hopkins' *Intruders*, despite Budd's status as arguably the world's leading expert on alien abductions. The advance was said to be one million dollars, *exclusive* of film rights. The book was *Communion*, and the author was Whitley Strieber. Whitley was already a well-known and successful writer before he published *Communion*, and was the author of a series of dark, contemporary horror stories, two of which—*The Hunger* and *The Wolfen*—had been made into films. *Communion* was an account of Whitley's encounters with the non-human intelligences he chooses to call "the visitors." The popular perception is that these "visitors" are extraterrestrial, but Whitley does not necessarily accept this hypothesis, any more than he accepts any attempt to label a phenomenon that may defy labeling altogether. Whitley first contacted Budd about his experiences, and seemed so distressed that Budd suggested he might want to see a psychiatrist, rather than a ufologist. Whitley eventually underwent a series of hypnotic regressions, first with Dr. Robert Naiman, and subsequently with Dr. Donald Klein, who was Director of Research at New York State Psychiatric Institute. Budd Hopkins was also present at many of these sessions, and although the two men were subsequently to fall out over a number of issues, *Communion*'s acknowledgments contain a heartfelt statement of gratitude for the help that Budd provided.

Communion ends with a statement from Dr. Klein in which he confirms that he found no evidence of any psychosis or personality disorder in Whitley, who—like many abductees I have come across—almost seems to *want* there to be a medical explanation, rather than have to accept the experiences at face value. *Communion* also includes the transcript of a lie detector test in which Whitley was assessed as being truthful in all responses relating to his experiences.

Communion is extremely well written, as one would expect from an experienced author. Perhaps in some ways it is rather *too* well written, because many of the descriptive passages are decidedly scary. I have come across a number of abductees who started to read *Communion*, but were unable to finish it. This is doubly unfortunate when one considers that Whitley's view of the phenomenon—once he had overcome his fear—is a positive one.

If *Communion* scares people, Whitley scares people too, and he certainly seems to have a reputation as a crank and an eccentric. I met him and his wife, Anne, when they paid a brief visit to London in 1996, and found this reputation to be unjustified. Whitley seemed a rather serious-minded man, and described some of his experiences in a thoughtful and methodical way. I had a suspicion that it was this very attitude that most disconcerted people, because he seemed so *casual* about the most bizarre experiences. Here was a man, I suspected, who would describe these encounters in the same sentence as an account of a trip to the mall, and might well use the same tone of voice. But then, if people's experiences are that frequent, and if they can come to terms with them, and accept them, maybe that is no bad thing.

Staring out from the front cover of *Communion* is the

image of what has been labeled by ufologists as a Grey: an egg-shaped head with its high domed forehead; a rather anonymous nose, and a slit for a mouth; upward-slanting, almond-shaped eyes: intelligent, hypnotic and enigmatic. This is the image that confronted the world not only from the book cover, but from hundreds of promotional posters. *Communion* became an international bestseller, and the image of the Grey began to inveigle its way into the public psyche.

1987 was a busy year. While the public was being hit by the double whammy of Budd and Whitley's books, the residents of Gulf Breeze in Florida were about to be hit by an unprecedented wave of UFO sightings. Hundreds of reports came in from this small coastal community, and these included numerous instances where there were multiple witness sightings. At the center of this wave were Ed and Frances Walters. For a time it looked as if this was going to be one of those intriguing cases that crops up every now and then, in which a cluster of sightings occurs in a particular location, for no apparent reason. Such *ufocals* include Warminster in Wiltshire, and more recently, Bonnybridge in Scotland. Gulf Breeze certainly attracted a great many ufologists and media, and a number of remarkable photographs and videos were taken. By any standard the Gulf Breeze affair is fascinating, not least because of an incident that occurred in 1990 when six American soldiers based in Germany deserted, traveled to Gulf Breeze, were arrested and then released amid rumors that their aim had been to expose an official UFO cover-up.

The Gulf Breeze story became even more interesting

when Ed Walters was revealed as an abductee. Apparently his experiences had started at the age of eleven, and continued over the years. Some of the features of Ed's experiences—the description of the entities, and his account of how he was incapacitated after having been struck by a blue beam of light—are common to a number of cases. Other aspects are more unusual, especially the fact that the aliens would refer to him as "Zehaas." Ed also recalled instances where something was clamped to his head, and somehow hooked up to an alien that was fitted with a similar device. During these occasions Ed would relive events from his past.

It seemed that abduction cases were becoming more and more exotic, with increasingly strange procedures being carried out. The experiences of people like Kathie Davis, Whitley Strieber and Ed Walters lay way beyond the boundaries of conventional knowledge, and tested to the very limit the resources of pioneering researchers such as Budd Hopkins. The problem was that most of these cases involved individuals, or groups of people all of whom had shared in the same experience. What was really needed was an abduction report with an independent witness who was not a participant in the experience itself. Better still, several such witnesses. And if one of those witnesses was somebody whom even the most cynical of skeptics would have to believe, then people would have to sit up and take notice. But that wasn't going to happen. Was it?

In April 1989 Budd Hopkins was contacted by a New York resident named Linda Cortile (a pseudonym), who believed that she was an abductee. She underwent hypnotic regression, which seemed to confirm what was now becoming an almost standard pattern involving a series of abductions (beginning in childhood) by small, gray beings, with procedures such as nasal implants that gave a medical perspective to the whole affair. This was by now becoming a familiar story to researchers like Budd. Linda's abductions seemed to have stopped, but she stayed in touch with Budd, availing herself of his network of abductees and their regular support group meetings.

In November 1989 Linda told Budd that she had had another abduction experience, parts of which (the onset of paralysis and the sighting of Greys in her bedroom) she consciously recalled. Under regression hypnosis she recalled being floated through the glass of her apartment window, floated up a blue beam of light into an oval-shaped craft (the case is sometimes referred to as the Manhattan Transfer), and having a medical procedure of some sort carried out on her, prior to being returned.

It was more than a year later in February 1991 when Budd received a letter purporting to be from two police officers. They told a story of how on November 30, 1989 they had been on patrol when they had seen a huge UFO near Brooklyn Bridge. To their amazement and horror, the craft had fired a beam of blue light at an apartment block, and they had then witnessed a woman dressed in a white nightdress floating up the beam, into the craft, accompanied by three much smaller figures. The police officers signed themselves Richard and Dan. This was dynamite, and Budd immediately realized that the event they referred to was undoubtedly Linda Cortile's November 1989 abduction.

A number of other strange facts were subsequently to emerge. It transpired that Richard and Dan were not police officers after all, but security personnel guarding a senior political figure. This figure apparently confirmed that he too had seen the abduction, although in a letter to Budd he referred to himself only as "The Third Man." Meanwhile, Richard and Dan seemed to be developing an unhealthy obsession with Linda, writing her unsolicited letters, turning up at her apartment, and following her around New York.

Budd then received a further letter that seemed to corroborate Linda's story from a woman who said that she had been driving over Brooklyn Bridge in the early hours of November 30, 1989 when she had seen the UFO, the blue beam, and the figure of a woman floating up the beam.

The identity of the "senior political figure" was eventually revealed by Richard and Dan. Although he has never confirmed the story, the man concerned was apparently Javier Pérez de Cuellar, the then Secretary General of the United Nations. The story became increasingly bizarre as time went by, and it wasn't long before the case was common knowledge among ufologists, who were split as to whether the case represented the breakthrough—in terms of corroboration—that they had hoped for, or whether it was an audacious and carefully constructed hoax. Budd Hopkins continued to research the story, and his 1996 book *Witnessed* deals with this most extraordinary of cases.

It is no surprise that the sorts of cases just outlined were to excite the attention of the media. The sheer number of television stations in America meant that access for those with minority views and offbeat ideas was much easier than it was in the United Kingdom. The turning point, of course, had been 1987, which had seen the publication of both *Intruders* and *Communion*. Budd Hopkins and Whitley Strieber found themselves in huge demand on the talk show circuit, appearing on some of the best-known programs in America, including *The Oprah Winfrey Show*, and *The Tonight Show* with Johnny Carson. Producers knew that abductees could give dramatic accounts of their encounters, and a number of them did so on various television shows. There are few things that make as gripping television as a good argument, and producers were always keen to get articulate skeptics such as Philip Klass on board, a tactic that certainly resulted in more than a few moments of fireworks.

Given the upsurge in public awareness of the abduction phenomenon, it was not surprising that Hollywood moguls began to give the matter some thought. *Communion* was made into a film, with Christopher Walken playing Whitley Strieber. Travis Walton's story was told in the film *Fire in the Sky*, and *Intruders* was made into a television mini-series. Aside from these "based on a true story" accounts, there were many other films and television series that incorporated the idea of alien abduction. *The X-Files,* although it features a range of paranormal events, is at its strongest when dealing with UFOs, government conspiracies and alien abduction. Even the 1996 smash hit *Independence Day*, while ostensibly a modern remake of the 1953 film *War of the Worlds*, incorporated an abductee who exacted a suitably spectacular revenge against his presumed captors. Soap operas were not immune either, and while British audiences scratched their heads and wondered if it was all some huge Hollywood joke, Americans recognized the imagery immediately when Fallon was abducted in a 1987 episode of *The Colbys*.

There is a Scottish rock band called CE IV (a Close Encounter of the Fourth Kind is a more official-sounding term for an abduction) who use imagery of Greys in their stage show, and material relating to abductions in the lyrics to some of their songs. A higher-profile use of such imagery was adopted by the pop group Babylon Zoo, who had someone dressed as a Grey on stage during performances of their 1996 UK number one hit, *Spaceman*.

Nowadays one can find books and magazines dealing with abductions, in any number of mainstream shops. Some of these magazines advertise products that range from the bizarre to the ridiculous, although the fact that there is a market for these items at all says much about just how all-pervasive the abduction concept has now become. Mugs and T-shirts are commonplace, but alongside advertisements offering regression, counseling and therapy in the aftermath of an abduction, one can find some more unusual items. How about an anti-abduction kit, to ward off the evil aliens? Or if you don't believe that's possible, then maybe you should consider an insurance policy to soften the blow. At least 900 people have taken out such a policy in Britain, while an incredible 150,000 Americans have subscribed. Before parting with your cash, you should bear in mind that most companies offering such policies insist that your abduction be officially acknowledged by a body such as the United Nations before they will pay out. Perhaps it is unfortunate for those pursuing a claim that Javier Pérez de Cuellar is no longer Secretary General!

It is a tradition that on Halloween, children dress up as ghosts, witches, vampires, or other such creatures, and go trick-or-treating. Nowadays in America some children wear masks in the likeness of Greys. People recognize the image.

An opinion poll carried out for *Newsweek* magazine and published in its July 8, 1996 edition found that forty-eight percent of Americans believe in UFOs, while twenty-nine percent believe that contact has been made with extraterrestrials.

Needless to say, there has been a less serious side to the escalating phenomenon, and magazines like the *Weekly World News* have run front-page headlines such as "Russians Shoot Down UFO—Angry aliens turn soldiers into stone" or the even more intriguing "Alien Backs Clinton! Space visitor tells Democrats how to rebuild U.S. economy!"

My favorite example of the extent to which alien abduction has infiltrated popular culture is the way that companies are using the concept in commercials. In an advertisement for Royal Insurance, a car is being floated up a beam of light toward a UFO. The occupants are on the ground, having presumably escaped the aliens, and one is saying to the other: "You'd better ring the Royal."

There is an almost identical American advertisement that was run by the telecommunications company AT&T, showing a golf buggy and its occupants being similarly drawn up a beam of light toward a flying saucer. The caption here was: "At a time like this, whose cellular phone would you rather own?"

Well-known companies such as these would hardly sanction a major advertising campaign unless they felt sure that the public would recognize the imagery.

Throughout this transformation of the alien abduction phenomenon from relative obscurity to high profile modern mystery, the skeptics had one question: where was the proof? Horror writers could tell their stories, modern artists might use hypnotic regression, but didn't this amount to little more than unsubstantiated accounts and theories that might be true, or then again might not. Where was the data? Where was the evidence? Where was the science? I will be expanding on these issues in chapter seven.

CHAPTER 5

The British Scene

The public perception of alien abduction is that it is a peculiarly American phenomenon. It is in fact global, but as always, public perceptions are influenced primarily by the media, who as far as abductions have been concerned have always concentrated on the American scene. There is no getting away from the fact that many of the key abduction incidents *have* occurred in America, and that many of the foremost researchers are American. Furthermore, the sheer number of television and radio shows in the States makes media access easier, making it more likely that "weird and wonderful" topics such as alien abduction will receive coverage. But there has been much activity in Britain, and by the time I began my investigations in 1991 I was inheriting a situation where events had been reported and investigated for nearly twenty years.

George King

Britain has had its fair share of contactees. Perhaps the most famous is London taxi driver George King, who claims that in 1954, after ten years of studying spiritual philosophy and yoga, he heard a voice that told him: "Prepare yourself; you are to become the voice of interplanetary parliament."

King claims that eight days later an Indian yogi taught him a technique that enabled him to contact an intelligence from Venus, who was given the name Aetherius. In 1956 he founded the Aetherius Society, which is still in existence today. King and his Society are controversial. There are many who believe him to be a fraud, and regard his group as cultists, albeit more of the eccentric than the dangerous type. Mainstream ufologists regard such groups with distaste, and when the British UFO Research Association (BUFORA) gave them a platform in 1996, at one of their regular London lectures, the move was received with mixed feelings, and the speaker was given a rough ride by one or two people in the audience. One of the prime sources of annoyance was that George King had awarded himself the title "Sir," to which he had no legitimate claim.

Jenny Randles is one of Britain's foremost ufologists and abduction researchers; her work suggests that abductions have been occurring in Britain for at least as long as UFO sightings. But she feels that in the early days these experiences were not followed up, because of a lack of awareness of the nature of the phenomenon. Although there had been a British abduction case dating from 1973, which was one of the first cases in the world to feature a combination of sexual examination and the production of human/alien hybrids, British ufologists in the seventies were generally skeptical about abductions. This mirrored the way in which American researchers such as Keyhoe regarded abductions with suspicion, believing they were fraudulent or mistaken claims that would serve only to discredit ufology as a whole.

I will now set out the details of a few British abductions, giving readers a brief tour through some of the key British cases.

The Aveley Abduction

An important British case occurred near Aveley in Essex in 1974, and the resultant investigation continued sporadically over the next few years. John and Susan Day, together with their three children, had been returning from a relative's house when they saw a UFO from their car. The initial sighting was fairly unexciting, and involved a blue light that appeared to be following them. Such accounts are always hard to assess, as the UFO is being observed through glass from a moving vehicle, and the resultant distortion makes accurate observations difficult. A little later their car radio began to malfunction, and actually gave off smoke. As the car drove around a bend it entered a patch of strange green mist, and was jolted about quite sharply. When they emerged from the mist they found they were much farther down the road than they had been the instant before. They arrived home, turned on their television, and were amazed to find that the evening's programs had finished. They had lost several hours of time. The case was investigated by Andy Collins and Barry King, and was the first case in Britain where hypnotic regression was used in an attempt to unlock the memories from the missing period of time. Under hypnosis the Days recalled being taken onto a craft by aliens, and subjected to medical investigations. But the aliens that the Days recalled were tall, fair humanoids, reminiscent of the sort of aliens reported by the contactees. Ufologists label such beings *Nordics*. As we shall see later in this chapter, when one compares the British and American case files there are some startling differences in the descriptions of the aliens reported as being responsible for abductions.

During one 1978 hypnosis session, at which Jenny Randles was present, Jenny asked Mrs. Day about the motives of the aliens, and had been given an explanation about how the aliens had genetically engineered the human race and were periodically checking up on us. What made this particularly interesting, however, was that Mrs. Day did not sound as if she was recalling a conversation she had had with the aliens during her abduction; she appeared to be relaying Jenny's questions to the aliens during the actual regression! This seemed unlikely, and was curiously reminiscent of the Maureen Puddy case from Australia, where the witness "saw" an alien while being driven to the site of a previous encounter, while the UFO researchers with her saw nothing. Was Mrs. Day simply confabulating, or were events occurring in an altogether different reality?

This case served only to confirm Jenny's growing concerns about the potential failings of regression hypnosis. In 1987, when American ufologists were falling over themselves to use hypnosis to investigate possible abductions, Jenny introduced a five-year moratorium on the use of hypnosis to investigate any cases reported to the British UFO Research Association, where she was Director of Investigations. It is interesting to speculate whether this bold move had a bearing on the considerable differences between British and American abductions that were to emerge in reports from the two countries.

Elsie Oakensen

On November 22, 1978 Elsie Oakensen had encountered a UFO on her way home from work, while driving along the A5 toward Towcester. At first she thought it was a low-flying aircraft that was about to crash. When she saw that the object was stationary, realization dawned that this was no plane. A closer look revealed that it was a gray, dumbbell-shaped craft, with two bright lights mounted on the underside of each end—one red and one green. As she neared her home in the village of Church Stowe the object was still in view, and the green light started to flash, as if it were signaling. Her car began to respond in an unusual way and at one stage Elsie recalls that it came to a halt, without her having applied the brake. Inexplicably, it appeared to be completely dark, even though it had been light seconds before. During this period a bright light shone onto the road from above. Suddenly it was light again, and the car was moving. When Elsie arrived home, she noticed that a journey that normally took her fifteen minutes had taken half an hour. The case was investigated by BUFORA, but although a few other witnesses were found, none seemed willing to speak out publicly, and the investigation was temporarily shelved. As it happens, a second witness *has* recently spoken out, and provided important corroboration of Elsie's account. Georgina Lourie was in a car with three friends at about the same time as Elsie's encounter. They too saw the lights, and their car engine faltered when the light came close. Finally, they saw the object fly away at incredible speed.

In 1979 Martin Keatman, one of BUFORA's regional coordinators, suggested the use of hypnosis to probe what might have occurred during the fifteen minutes of missing time. Elsie agreed, and a session was conducted by Graham Phillips and Andy Collins; while regressed, Elsie recalled a bright light coming toward her, followed by the sighting of two ghost-like entities.

The encounter had a transformative effect on Elsie's life, and she became involved in areas that had previously been of no concern to her. These included interest in psychic events, and the development of real ability as a spiritual healer. Whatever happened to Elsie Oakensen, she regards it as a gift.

Lynda Jones

On a summer's evening in 1979, in Didsbury, near Manchester, Lynda Jones was out walking with her young son and daughter. Suddenly they saw an object hurtling toward them from the sky, and thinking an aircraft was about to crash, they crouched down in the field. When the object had passed overhead they ran up a hill to get a better view. To their amazement they saw a crescent-shaped UFO. Lynda was extremely curious, and felt drawn to it. Time seemed to stand still for her, and she walked toward the brightly lit object. Eventually, prompted by the shouting of her daughter, all three ran off.

When they got home, one and a half hours of time was unaccounted for, and the following morning Lynda's husband Trevor noticed that she had red marks, similar to blistering sunburn, around her eyes.

Under hypnotic regression Lynda recalls seeing a figure in the field, followed by a floating sensation. She then recalls being in a room with three tall figures with black hair, jumpsuits and large, black, almond-shaped eyes. The figures were shining a bright light in her face. She does not know how to evaluate this incident, which she freely admits was totally outside her knowledge or experience.

This investigation was led by Harry Harris, an energetic Manchester-based lawyer who is a strong supporter of hypnosis as an investigative technique. Harry has used a considerable amount of his own money to pay for qualified hypnotherapists to help with the investigation of a number of his cases.

Bob Taylor

It is debatable whether this case is an abduction. But although there was no recollection of any entities, the story is nevertheless significant because of the apparently hostile action of a UFO. On November 9, 1979 forester

Bob Taylor was walking his dog through Dechmont woods near Livingston, in Scotland, when he saw a huge, domed UFO hovering silently about two hundred and fifty feet above the ground. Two small, round balls detached from the craft and moved rapidly toward the startled forester. The objects had spikes on them, and looked rather like old naval mines. The balls attached themselves to Taylor's trousers, cutting him, and he recalls an acrid smell, which was his last memory before passing out. When he came to, he recalled the sound of an object moving off at high speed. When he returned to his truck, it would not start. He reported the incident to the police, who went to the site of the encounter, and concluded that an unknown object weighing several tons *had* been in the clearing.

Taylor's employers and friends—one of whom I met quite by chance a few years ago—were unanimous in their view that he was a straightforward man with absolutely no motive whatsoever for fraud.

Alan Godfrey

One of the most famous abduction cases ever to have occurred in Britain happened on November 28, 1980 and involved a police constable based in the Yorkshire town of Todmorden. Godfrey was nearing the end of his night shift when he received an order to investigate reports that some cows had escaped from a farmer's field and were getting into private gardens. Godfrey drove toward the area, and as he rounded a particular bend in the road, saw what he at first took to be a bus in the road ahead of him. Upon closer inspection he saw that it was a diamond-shaped

UFO hovering some five feet above the road. Godfrey attempted to contact his headquarters, but as so often happens when in close proximity to UFOs, the equipment would not function properly. Godfrey decided to sketch the object on his notebook, and after having done so found himself suddenly and mysteriously about a hundred yards farther down the road. The UFO had vanished. Godfrey returned to his police station, picked up a colleague, and returned to the site of his encounter. They both noticed that the road was considerably drier at the point where Godfrey had seen the craft. What puzzled Godfrey even more was that when he had returned to the police station, it had been much later than he thought.

Under hypnosis—again, organized and paid for by Harry Harris—an abduction story emerged. Apparently Godfrey had got out of the car for a better look at the object, but had suddenly seen a bright light emanating from the craft. He had been frightened, and had tried to drive off, only to find that his car would not start. He felt he was being restrained, and then suddenly recalled being inside the craft, where he saw a tall, bearded humanoid figure called Joseph, together with smaller creatures who Joseph explained were robots. At one stage Godfrey even described seeing a dog on the craft: the sighting of animals inside UFOs is highly unusual, although an American abductee—Judy Doraty— recalls seeing a cow being floated onto a craft, prior to the unfortunate animal being dissected.

Godfrey said that he was placed on a table of some sort, but despite recollections of a bright light, and some sort of machinery, he was unable to remember any further details. Godfrey had been extremely distressed during his first hypnosis session, and during the second told the investigators that he was "not supposed" to tell them about the machinery.

Alfred Burtoo

This fascinating case is probably better classified as a contact experience rather than an abduction, but is important, if only to show that the contactees have not gone away altogether, even if some ufologists think they have. On August 12, 1983 Alfred Burtoo was night fishing on the Basingstoke Canal near Aldershot, Hampshire, when he saw a bright light approaching. His dog started to growl, and all of a sudden he saw two entities walking toward him. They were about four feet tall, dressed in green one-piece coveralls, and their heads were covered by some sort of helmet. They beckoned to him to follow them, which he proceeded to do. Shortly thereafter, on the towpath, he encountered a landed UFO, and followed the beings inside. After some time inside the craft, during which Burtoo had been observing his surroundings, he was told to stand under a light, which he duly did. He was asked his age, and told them that he was seventy-seven. To his disappointment, he was told that he was too old and infirm for their purposes, and then left the craft. Returning to his fishing equipment, he subsequently heard a whining noise, and saw the craft take off and fly away at high speed. Incredibly, perhaps, he calmly picked up his equipment and carried on fishing! Burtoo died in 1986, never once changing his story.

The Ilkley Moor Abduction

On December 1, 1987 a former police officer, Philip Spencer, was walking on Ilkley Moor in Yorkshire, with the intention of taking some photographs. He heard a low humming sound, and then saw a strange entity on the moor. He took a photograph of the creature, and proceeded to follow it. It led him to a UFO that seemed to consist of two separate saucers, somehow joined together. The entity got into the craft, which took off, accelerating away at high speed. When Spencer returned home he found that he had lost two hours of time, and during a subsequent hypnotic regression he recalled being taken into the craft and placed on a table by creatures with pointed ears. Like a number of other abductees, he was then shown a film that contained apocalyptic images. He was also shown a second film, the contents of which he was told not to reveal.

James and Pamela Millen

During September 1990 the Millens went camping in Dorset, and one night they awoke just before 3 a.m. Upon going outside, James noticed several orange balls of light that seemed to be dancing in the air over an adjacent field. He pointed these out to Pamela, and the couple watched, mesmerized, as the balls of light moved silently around in the sky. After a couple of minutes James decided to go and get his camera, but what happened next remains unclear. Pamela recalls that James suddenly appeared and commented on a pile of cigarette ends at Pamela's feet, which seemed odd, as he had only been away for a minute or so. But when they looked at their watches it was now

5:40 a.m. Over two and a half hours of time was unaccounted for.

James subsequently had recollections of being in a circular, white room, lying on his back on something cold, and seeing figures dressed in robes rather like those worn by the Ku Klux Klan.

James has no idea what happened to him, but believes that it was something "beneficial."

The few cases discussed so far are only the tip of the iceberg, but serve to counter any suggestion that alien abduction is solely an American phenomenon. There are several hundred British cases on record, and several such cases that I have investigated will be covered in depth in Part 2 of this book. Then there are the countless abductions that go unreported through fear of ridicule or lack of awareness of whom to speak to about such matters. During my official investigations at the Ministry I came across many people who would never have dreamed of discussing the matter with a ufologist.

Witness-Led Investigation

The moratorium on the use of hypnosis introduced within BUFORA in 1987 had left a vacuum in British research into cases of alien abduction, although there were other UFO groups and researchers who used the technique. Into this vacuum was to step one of the most innovative thinkers in British ufology, the late Ken Phillips, a BUFORA investigator whose tragic and untimely death in July 1996 deprived the world of abduction research of one of its finest minds.

Between 1981 and 1992 Ken Phillips had run an in-depth analysis of over one hundred close encounter wit-nesses. This study, known as the Anamnesis Project, was aimed at seeing whether there were any common features in the background and lives of the abductees that might have a bearing on their experiences. The study had been drawn up in consultation with Dr. Alex Keul from the Uni-versity of Salzburg in Austria. There were some in-teresting findings, including a high incidence of claimed psychic abilities among abductees, together with the fact that they generally held less prestigious jobs than might have been expected, given their intelligence and ability.

The significance of the Anamnesis Project was that it moved the focus of the investigation on to its most impor-tant feature—the witness. Many of the personalities in-volved in abduction research are strong characters who may unwittingly have an effect on the information that emerges. As I explained in chapter three, this is particu-larly true where regression hypnosis is concerned, and skeptics often allege that the investigator—who takes on a very powerful and central position in a subject's aware-ness during hypnosis—may, through the use of leading questions, arrive at an account that matches his or her own views, rather than one that reflects what actually took place. Mindful of such criticisms, and wary of the situa-tion that was developing in America, where abductions were almost invariably being portrayed as negative, Ken Phillips decided to set up a Witness Support Group where the whole concept of investigation would take a backseat. Instead, the abductees would simply come together for mutual support, to discuss their experiences and help as many people as possible to come to terms with them.

The group first met in February 1990, and was a great success. It transpired that many of the abductees had developed interests in psychic matters, and had discovered talents in this area. Others had suddenly become creative, and had started writing songs and poetry. But the most important benefit to emerge from the group's meetings was the simple realization that its members were not, after all, alone in having had their strange experiences.

Differences Between British and American Abduction Cases

Throughout this chapter it will have become apparent that there have been a number of differences in the sorts of reports to have been documented in Britain and in America. Jenny Randles has made particular attempts to draw together British abduction reports, to see what patterns emerge. An in-depth study of forty-three potential abductees threw up the curious fact that reports of Greys in Britain were comparatively rare. Nearly half of the cases involved abductors who were reported to be essentially human in appearance, and many of the others were of the Nordic type. It also transpired that the number of witnesses who wanted any kind of publicity was smaller than might have been expected from the American experience. Of the forty-three witnesses in Jenny Randles' study, only three were prepared to go public. It is dangerous to draw firm conclusions from analysis of so few people, but at the very least, some fascinating questions are posed by these results.

As well as the physical differences in entities, we have seen differences in the way in which these cases are investigated, and I cannot help but think there must be some significance in this, because the alternative (which seems unlikely) is that extraterrestrials respect national boundaries and undertake varying activity according to their geographical location. To a certain extent the way in which an experience is perceived and interpreted must depend upon the methodology used to investigate the case, and the belief systems of abductees and investigators alike. This is an area that needs further research, especially to clarify the situation elsewhere in the world and assess whether any geographical patterns can be established. There are plenty of abduction reports emerging from countries like Brazil and South Africa, but often this simply reflects the tenacity of investigators in particular countries, and their abilities at tracking down cases. What is needed is a far wider effort, targeting countries that have not previously been looked at and published.

The Current Situation

Lately there has been something of a backlash against BUFORA's restraint. I am finding that increasing numbers of British abduction researchers are becoming more bullish about the use of hypnosis, and more upbeat about its effects. Investigators such as Tony Dodd, Roy Lake, Harry Harris, Linda Taylor and Eric Morris are vigorous supporters of the technique. Magazines dealing with UFOs and abductions have proliferated over the last year or so, and a glance at their classified columns reveals the sheer number of investigators and therapists offering regression hypnosis to abductees.

Harry Harris recently told me that while everyone recognized there were pitfalls involved in hypnosis, it remains the best technique in investigators' armories. The challenge to its critics was therefore to find an alternative. I sense that views are polarizing, and that the debate on the issue is about to become extremely heated. There is nothing wrong with vigorous debate, provided that the witnesses themselves are not caught up in the fray, finding themselves being used as pawns in a ufological power struggle.

CHAPTER 6

Textbook Abductions

Alien abductions seem to occur all around the world, and are surprisingly similar in structure. Almost identical details have been reported—quite independently—from a vast number of abductees. While some may have been familiar with these details simply because they had come across them in the UFO literature, or in the media, the commonality in descriptions was noticed by researchers long before the material was widely available in the public domain. This strongly suggests that a real phenomenon is at work; it may be perceived in different ways, according to the culture and belief of the witness, but the phenomenon is essentially a fairly standard one.

To familiarize ourselves with the multifaceted nature of the abduction phenomenon, I have constructed two fictitious abductions. In doing so, I have drawn all the commonly reported elements together in a way that does not always happen in actual cases. In this fashion, the two accounts will serve as blueprints for the entire phenomenon. Without exception, however, the events described in these fictitious accounts mirror events reported in cases that either I or other abduction researchers have investigated. Although many of the features are self-explanatory, I will make some bracketed observations as we reach certain key points. Most of the theorizing, however, will not be covered until later.

Road Abduction

The business meeting had gone on well into the evening, and although John Smith was now driving home, he knew he would not arrive for some time yet. As he was recently divorced, he mused that the lateness of his arrival would not result in the sort of argument that had been a common feature of his life until recently. *[I have noticed that abductees seem generally to be less likely to be in a stable relationship than one might expect with a control sample. This might be due to resentment—conscious or subconscious—on the part of the abductees that their partners are unable to prevent the abductions.]* He glanced at his watch, and noticed that the time was just after four in the morning. Stifling a yawn, he rolled down the window to let in a little air, mindful of the dangers of driving when drowsy. As an additional means of staying alert he flicked the switch on his car radio, which was pre-set to a channel devoted to rock music.

The occasional car had passed the other way, but the road was now completely clear. This was a little unusual, because despite the late hour, this was a fairly major route, and was particularly favored by long distance truck drivers. *[It is a curious feature of road abductions that no other cars are seen immediately prior to the event, even on a busy road, and that no cars pass by while events are occurring. This implies either that an extraordinary degree of control is exerted not just on the abductee, but also on the surrounding environment, or that events do not take place in the physically observable universe. This is possibly connected with the way in which other sensory inputs seem to diminish prior to an abduction, typified by a sudden quiet. As previously mentioned, this otherworldliness has been called "The Oz Factor," and might imply that events occur in another reality altogether.]*

At first, John thought that the lights in his rearview mirror belonged to another vehicle, and he was annoyed that someone was driving so close behind him. If the other driver was in such a hurry, why not overtake, rather than sitting on his bumper? The road was straight and clear, after all. All the same, he felt vaguely uneasy, and after a few minutes, with the other vehicle showing no signs of overtaking, he decided to put his foot down. None of this seemed to have any effect, and he had a sudden terrible thought that he was being followed by a psychopathic truck driver, as had happened in the early Steven Spielberg film, *Duel*.

All of a sudden, the radio began to crackle, and the song that had been playing disappeared into a sea of static. *[Interference with a car radio or household electrical equipment is frequently reported in association with UFO encounters.]* He turned the radio off, not wanting this additional distraction while he contemplated what to do about the vehicle behind him. Glancing again at his rearview mirror he noticed that the lights did not appear to be in the standard configuration, or indeed in the normal color scheme. Just as he wondered whether this was some sort of customized vehicle, he noticed something that sent a sudden chill through him; the lights were higher than they should have been. The pursuing vehicle was not on the road at all, but *over* it. Fully alert now, he did some rapid thinking. Despite being cocooned in his vehicle, he would have been able to hear the noise of a helicopter or an aircraft. Might it be some sort of glider, or small plane? Was it in trouble? He had heard stories of aircraft in difficulties landing on roads. Maybe he should pull over onto the hard shoulder. It was then that his car engine spluttered and died. *[Vehicle interference is another feature often reported in relation to UFO encounters and road abductions. It is not known whether this is something done deliberately by the UFOs, or whether it results from electromagnetic interference from the UFO's propulsion system.]* As his car slowed and began to slide off the road it wasn't panic that overwhelmed him, but a strange feeling, like sleepiness . . . only different.

John blinked hard, and suppressed a brief moment of nausea. Had he nearly drifted off to sleep, he wondered. He felt disoriented, but before he could decide whether or not he should pull over he saw the sign indicating that the turn-off for his home town was coming up. As he was so close to home, he decided to press on, but it was then that he began to think something was wrong. He was sure that he hadn't gotten this far with his journey, and he couldn't remember having passed many of the landmarks on what was now becoming a depressingly familiar late night drive. As he pondered this mystery, trying to replay the journey in his mind, he had a momentary flashback. What had happened? Something to do with a light? The moment the word "light" flashed into his awareness, he realized what else had been troubling him. It was getting light. He told himself that it couldn't be, but a glance at his watch told a different story. It was nearly six in the morning. His journey had taken almost two hours longer than it should have. What the hell was going on? *[So-called "missing time" is the most common indicator that an abduction may have occurred, even if one is not consciously recalled at the time.]*

Two weeks after these events, John found himself in a position that he had never in his life anticipated: he was flat on his back on a couch, in the office of a psychotherapist. John was having difficulty sleeping, and when he did get to sleep he had been haunted by a terrifying series of nightmares. *[Memories of abductions—and indeed any forgotten trauma—can sometimes surface in a disguised form, in dreams.]* He couldn't remember many of them but there was a recurring image of a large insect with horrific eyes, staring at him. *[This mirrors a phenomenon known as "screen memory," when abductees see an alien, but perceive it as something more familiar. Because of the fact that the most striking feature of aliens recalled by abductees is the large eyes, these screen memories often involve creatures with similarly striking eyes, such as cats or owls. It is not known whether this is due to the human mind trying to match the entity with a creature that it can identify, or whether it is something deliberately induced by the aliens to hide their appearance.]*

The initial consultation had produced a few surprises, not least when the therapist had quizzed him at length about his journey of two weeks ago. While he accepted that the whole affair was bizarre, and that his troubles had started thereafter, he failed to see any significance in the events, and indeed nothing particularly useful had come out of this initial session. When rolling up his sleeves he had noticed a strange scar, but thought nothing of it. *[Many abductees report strange scars that they cannot account for, and it has been suggested that they are scars resulting from medical procedures carried out during abductions.]* To his surprise the therapist had then suggested the use of hypnosis, to aid relaxation, and perhaps to unlock a few memories.

John's second session had begun with a few reassurances about hypnosis. He was told to ignore any stage hypnosis he had seen. Hypnosis carried out by a trained hypnotherapist was simply a way of focusing the mind; it wasn't a trance, and he would be conscious of his surroundings at all times. Reassured by this he had agreed to be hypnotically regressed to the night two weeks ago. He felt very relaxed, but not at all unaware of what was going on. Like all those being professionally hypnotized for the first time, he wondered vaguely if he was properly hypnotized at all.

Three and a half hours later John Smith was back in his own home, his mind in an absolute turmoil. Far from solving his problems, the regression had uncovered a story so bizarre that he was beginning to question his sanity. Even the therapist had been at a loss for words, although apparently some of her colleagues had come across similar cases; she would be contacting them prior to the next session, to see if they had any idea as to how to handle such a case. During the session John had recalled the incident with the lights following his car, the radio cutting out, and then the engine dying on him. And then . . .

. . . The engine had cut out, but the car shouldn't be moving in this way. It was sliding diagonally off the road, as if it were on tracks. It was actually a very smooth movement. The car came to a halt, and he was aware that he was at the side of the road. His hands were gripping the steering wheel, but he didn't feel the fear that part of him told himself he should be experiencing. Something was badly wrong, and there was an eerie feel to the whole situation. Aware of light around him, he got out of the car. Intellectually, he knew that the most rational explanation was that he had been forced off the road by another car, which was now parked up ahead of him. But on a more intuitive level he knew that something else was happening here. There was something on the road ahead of him all right, and it was very brightly lit. But it wasn't a car.

He caught sight of movement, and saw figures against

the light. He put his hand above his eyes, trying to block out some of the light; straining for a better view. For some reason a thought implanted itself into his consciousness, in a way that was unlike any thought process he had ever experienced before. He felt he was looking at some road-works, surrounded by warning lights. Marshals were directing him around the hazard, and that's what he was supposed to remember about the experience. He reassessed the situation, realizing that this was *not* the explanation for what was happening. He felt a surge of anger. Something was trying to trick him, but he wasn't going allow that. He told himself over and over again that it wasn't roadworks he was seeing. *[Like screen memories of the creatures themselves, it is not entirely clear whether such tricks are induced by the abductors in an attempt to mask their activities, or are constructed by our own minds, in an attempt to make sense of information that conforms to nothing within our memory or knowledge. If we have no frame of reference, perhaps our minds pick the closest match.]* He had been edging closer to the light source, not entirely sure as to whether or not this was at his own volition. He felt as if he were being compelled somehow, and was moving as if under some sort of hypnotic control. As he looked into the light he likened himself to a rabbit, caught in the headlights of an approaching car.

He was close enough now to make out more details. It was . . . well, what he could only describe as a flying saucer. One look at it convinced him that it could only be from another world. So, he thought to himself, those "UFO nuts" were right after all. But somehow, despite the sheer wonder that he felt, the most practical of thoughts popped into his head. Why, he wondered, had it stopped in the middle of the road? If another car was to come along . . . well, what would happen then?

He was now no more than ten feet from the craft. It was roughly saucer-shaped, with various blue, white and red lights mounted on the outside. Some were flashing, and some appeared to be rotating around the outside of the hull. It was, he supposed, about twenty yards in diameter. But it was the figures that held his attention. There were three of them, and they were about three—no, three and a half feet tall. They were a sort of pale gray color, naked, with no hair, and no visible sexual organs of any sort. Although humanoid in appearance, they were clearly alien. Their arms, legs and torsos were spindly, and their heads were shaped like an inverted drop of water; a little bit like an egg maybe, but slightly more exaggerated. The nose was tiny, and the mouth was just a slit, but it was the huge eyes that commanded his attention. The eyes dominated the entire face, and indeed were the most noticeable feature that the beings possessed. They were almond-shaped, and slanted upward. They were jet black, although they reflected the lights from the craft.

He felt compelled to look into the eyes of one being. He felt himself mesmerized, and unable to look away. He was being told—somehow—to follow the creature. This offended his dignity, and he fought the compulsion. At this, the being produced (he didn't see from where) a short, black rod, which was about a foot in length, with a bright blue/white light at the end. The being pointed this at him, and it fired a beam of blue light into his body. *[The similarity between these devices and the magic wands of our folklore is fascinating. Could our legends be more closely based on fact than we might suppose?]* He didn't exactly feel pain, but he certainly lost any will or ability to resist, and he felt himself walking toward an open hatch in the side of the craft, while the three beings walked at his side, prodding him along. There was a slight ramp, and once inside, the first thing he noticed was that it was a little darker inside, now they were out of the glare of the dazzling external lights. He heard a faint noise at his back, which he assumed was the hatch closing behind them. He wanted to turn around to look, but was unable to move his head.

The inside of the craft was cramped, and he had to stoop. The walls were white and featureless, and there were two or three doors on the right, as he was led clockwise along what he assumed was a corridor that ran in a circle around the edge of the craft. He tried to peer through the doors as he passed, but was unable to see much. But through one he saw—or thought he saw—a man in uniform standing at some sort of control panel. *[There have been a number of such reports involving humans carrying out tasks on UFOs, and sometimes even helping with medical procedures on abductees. While some feel these "collaborators" are simply other abductees acting under compulsion, conspiracy theorists speculate that they are government or military personnel, actively liaising with the aliens, and perhaps taking part in some sort of bizarre exchange program.]* They approached another door, and this time he was ushered through.

It was a round, white room, and in the middle was what appeared to be a table. He was full of questions, but was somehow unable to think straight, let alone speak. He was led to the table, and maneuvered into a situation where he was standing by the side of it. He then received what he could only describe as a compulsion to take his clothes off. He heard no voice, and the word "telepathy" was hopelessly inadequate to describe the sensation he had, but he knew that it was an instruction that had come from them. He removed his clothes without thought of resistance, and following another instruction, lay on the table, on his back. He could somehow sense the beings' feelings. They were businesslike; impatient, almost. This was not some great meeting between two cultures. He was the laboratory rat in the maze, being handled by the scientists.

He felt a presence at both sides, and assumed the beings were there, although he was unable to see them. After a few seconds, one did come into view, and stood at the foot of the table. He had what appeared to be a tray in his hands, with various medical-looking instruments on it. He tried to cry out, but felt a voice in his head: "Don't worry. There's nothing to be afraid of. This won't take long."

All of a sudden, one of the beings at his side made a rapid movement, and inserted a long probe of some sort right up his nose, causing him to scream in agony. There was a curious moment of stillness, as if the creatures had been shocked and stunned by his scream. He felt as if he had achieved a small victory, and somehow gained the initiative now that they were reacting to him, and not vice versa. The pain subsided, and as the probe was withdrawn, he saw that a tiny metallic device was visible on the tip of the probe. As the tiny object was dropped into a tray, he wondered whether this device had been removed from his nose. But if so, what was it, and why had it been there in the first place? Might these explain the nosebleeds that had been troubling him recently? *[It is speculated that such implants are tiny locators, enabling the abductors to find people who have been previously abducted and fitted with such devices. Sometimes these devices will irritate the nasal cavity, and cause nosebleeds. If implants exist, and if they are tracking devices of some sort, it suggests that abductions are an ongoing process in an individual's life, rather than a once-only experience. This is a highly controversial area, and despite claims that some of these implants have been recovered, and despite ufologists having produced some strange-looking devices, we have yet to see anything that the world of science can prove is not of this Earth. As I write this, the American researcher Derrel Sims has produced a number of anomalous objects that have apparently been removed from abductees by doctors. As far as the medical profession as a whole is concerned, however, the jury is still out.]*

Following this rather unpleasant operation, his body was then bathed in a warm light, which seemed to pass up and down his torso, penetrating to his very bones. Suddenly, he was aware that he was feeling sexually aroused, which made him feel first amused, and then embarrassed. He was aware of some pressure around his genitals, followed by a feeling that something was being attached to his penis, which was, by now, erect. Sexual imagery seemed to flash in front of him somehow, as if erotic pictures were being fed directly into his mind. He seemed to have no control over this, and felt it was something that was being induced in him. He ejaculated, and then felt a momentary pain as the device was removed from his genitals. He was upset and angry, and felt that he was simply being abused, both physically and mentally. He felt totally humiliated, and was crying. At this point, he was aware of the presence of another being at his side. The total paralysis that had gripped him during his recent ordeal seemed to diminish, and he regained some control over his own movements. He still felt constrained, but he could at least move his head a little. More importantly, he felt that his thought processes were becoming clearer. Previously, he had been in an almost dreamlike state, but now he was much more self-aware.

The being at his side was different from the others. It was taller, and even though there were no indicators of gender, he seemed to know that this being was a female. *[Abductees often sense their captors as being of the opposite sex, leading some to speculate whether some form of subtle sexual attraction might be used as a form of control during the experience. Further research into this, and into the perceptions of gay abductees might help to resolve this speculation.]* Incredibly, there was something familiar about this alien. Even though he knew that he was embroiled in an experience so bizarre that he was having difficulty believing it, he felt that he knew his abductor. Indeed, a sudden and inexplicable feeling of love and warmth grew within him, and at this point he would have given his life for this creature, had she asked it. It was not that he felt sexually attracted to her. But there was love, of a sort, and unlike his earlier physical arousal, this seemed genuine, although logic told him that it probably wasn't.

The pain and degradation was over, and now he was with this other creature. The creatures that had abducted him were nothing, he felt. They were drones, and had simply carried out a set piece. Now, he was in the presence of a truly wonderful creature, and as he looked into her huge eyes, he sensed both intelligence and compassion.

All of a sudden he felt a series of images appearing in his mind, as if a projector was spewing forth pictures at an incredible rate. They were flashing through his awareness so quickly he could hardly focus on any individual scene. The sense of it was clear though. He was seeing apocalyptic images of the Earth, involving floods, explosions, and death. The final image was of the Earth as a dead planet, with a sulfurous atmosphere and an oily, black sea. *[The showing of apocalyptic images is another common feature of abductions, and has led some researchers to suggest that the aliens are trying to educate us, warning us of the dangers we face. If so, this would be an interesting parallel with the rather more open warnings given to some of the contactees.]* He felt a sense of incredible sadness, and deep loss, but as suddenly as this feeling grew, it diminished, as if his thoughts were being taken from him. He was aware of the being staring at him, and sensed that she was pleased with him.

"We will meet again soon, but for now, you will remember nothing." The voice in his head was beautiful and hypnotic, but the whole experience had offended his sense of pride, and he told himself that he *would* remember. There was something strange about the voice in his mind; something deeply compelling. He felt himself locked in a battle of wills with the creature, but he could not bring

himself to hate her. It was as if they were old friends, although rivals. He tried to shut out the voice, and repeated
to himself over and over again that he would remember:
he would remember—this time. As the words "this time"
occurred to him he felt a profound sense of shock, but before he could ponder this further, he felt himself being half
dragged, and half carried out of the room. He was in a
daze, and began to lose the clarity of thought that he had
experienced after the procedure that had just been carried
out. He felt as if he were in a dream, or maybe drunk.
There was a blast of cold air as he emerged from the craft
into the cool night. He was vaguely aware of being maneuvered back into his car, but he could not say for sure
how the vehicle was started, or how, exactly, he regained
control of his physical and mental faculties.

John sat in his armchair. He was shaking, and crying,
and all the while he was trying to make sense of the material that had emerged under hypnosis. What disturbed him
most was the feeling that far from having undergone some
wild new experience, he had actually done no more than
unearth something that had been happening to him all
his life.

Bedroom Abduction

Sandra Hughes knew that she was an abductee. Unlike
many, whose memories of these experiences had only
emerged under regression hypnosis, Sandra could recall
significant portions of her experiences unaided. Sometimes
she felt this was a curse, given the intrusiveness of the phenomenon, and the pain. But most of the time she was
grateful that she was not tormented by the confusion and

uncertainty that dogged those with no conscious memories of their abductions. She had met other abductees, and found that they were, generally speaking, plagued with doubts about their experiences. There were one or two who wanted to believe, and seemed almost pleased to be classified as abductees. Most, however, were looking for more prosaic explanations. They would probably be ecstatic if it could be shown that their experiences were psychological. The last thing in the world they wanted was for their cranky, quirky memories to represent a literal truth. Regression hypnosis gave them their get-out clause. Intellectually, most were aware of its limitations, and could allow themselves to believe that their memories of abduction events might have been confabulated, or accidentally induced by a credulous hypnotist who had unwittingly led the witness to recall something that matched the therapist's own beliefs.

Sandra's conscious recollections freed her from this internal debate, and she felt this helped. There was no doubt for her, and no debating the psychological complexities of hypnosis and belief. She was frequently abducted by aliens, and she generally remembered enough of the experience to be sure that it occurred in the *physical*—and not the psychological—universe. Although she had undergone regression hypnosis, she had only begun the sessions recently, in an attempt to come to a better understanding of precisely what was happening. The material that emerged under hypnosis served only to confirm her conscious recollections, and this gave her confidence in the technique.

Sandra had returned from a therapy session a few minutes ago, and was trying to make sense of the information that had emerged. She had consciously recalled the beings coming to her bedroom, and it was this that persuaded her to be regressed again, as she felt sure that something particularly significant had occurred. Now, after hypnotic regression, she was convinced that last night's spectacular encounter was a turning point in her relationship with the visitors, and one that had put a lot of other episodes of her life into sharper focus.

It had started at ten past ten at night. Sandra and her boyfriend, Dave, had gone to bed early. She had drifted off to sleep, but awoke suddenly, with a start. Although fully conscious, she felt pressure on her chest, as if something was holding her down. She tried to move, but was unable to. A paralysis seemed to have overtaken her entire body, except her eyes. She fought down a feeling of panic, and tried to slow her breathing. She was familiar with lucid dreams and false awakenings, but knew that this was neither. She was also conversant with the theories about sleep paralysis, but she was well aware of what was taking place. It had, after all, happened before.

She glanced toward the foot of her bed, but when she saw the three figures standing motionless there she gave no more than an involuntary start; the whole scenario was, after all, depressingly familiar. They were about three and a half feet tall, and light gray in color. They were all dressed in what appeared to be one-piece uniforms of some sort, but with no insignia. Their limbs were frail looking, and their heads disproportionately large. The dominant feature on the creatures' humanoid faces were the eyes, which were jet black, and extremely large. Whichever way Sandra looked, those eyes seemed somehow to follow her. There was no escaping them and the hypnotic power that they exerted over her.

Suddenly the room lit up, filled somehow with a bright light that had no apparent source. The light was white, with perhaps just a hint of blue. Aware that the paralysis seemed to be wearing off, she tried to rouse Dave, shaking him violently, and shouting at him to wake up. *[In a scenario where somebody is abducted from any situation where others are present, the others seem, somehow, to be "switched off."]* It was to no avail, and in frustration, she threw her pillow toward the figures at the foot of her bed. At this, she was suddenly overtaken by a feeling of extreme lethargy, and her body went limp. She was aware of hands reaching down for her, and helping her to her feet. She was taken from her bed and led toward the window. She felt groggy, and although she was walking, would have fallen were it not for the fact that her abductors were supporting her. There was a large window in her bedroom, which was several stories up a block of apartments. She had a sudden sensation of floating, and screamed as she moved toward the glass. Both she and the three creatures somehow passed through the solid glass, and the sensation was far from pleasant. Sandra found this part of her experience the most frightening of all, and was overtaken by a wave of panic as her mind filled with images of getting stuck halfway through the window, with glass lacerating her torso. She fought down a wave of nausea, as their slow progress through the solid matter continued. It felt as if she was struggling through quicksand, and she made a mental note to try and sleep with the window open, to save herself this ordeal. She wanted to ask the creatures why they couldn't just open the window, or even ask her to do so. She'd be happy to help if it meant avoiding that horrendous sensation. But, speech wouldn't come, and she was unable to interact with her captors in any way.

The next instant, she was aware of a sudden lifting sensation, as if she were being sucked upward through a pipe. She glanced downward, and caught sight of the houses below her. She wondered if anyone might see her, and whether if they did they would report it to someone. This was happening over a major city; why couldn't someone see all this, and do something to help her? Her feelings of frustration and anger were overtaken by panic as she realized how high above the ground she was. What if the aliens made a mistake? What if she fell? *[Many abductees are afraid of heights, and will not travel in elevators. These phobias may well stem from such experiences within an abduction.]*

There was an object just above her, and she tried to look up to see. She knew it was a craft, of course, but wanted to get a clear look. All she could see was the light from the beam that she and the three beings were traveling up. She felt suddenly disoriented, and the very next moment she was inside the craft, with no clear memory of exactly how she had entered into the structure. *[So-called "doorway amnesia" is a fascinating theme running through almost all abduction accounts. Nobody is entirely sure why memories of the actual moment of entry should be so hard to recall, even under regression hypnosis. It seems clear that this is an important point, the resolution of which might take us considerably nearer an explanation of the true nature and purpose of the abduction phenomenon.]*

She was lying on what she assumed to be a table of some sort, and although there were no physical restraints, she was unable to move any part of her body except her head. As she glanced around, she saw the three creatures who had abducted her scurrying around at the end of the table. The room she was in was circular, or perhaps oval, and was largely featureless, except for some strange hieroglyphic symbols on the white wall. She made a particular note of these, and promised herself that she would remember them. There was no confusion in her mind; she knew she had been abducted by extraterrestrials, and taken onto a UFO. She knew this had happened before, and that it would happen again. She also felt that she was locked in some sort of battle of wills with her abductors; she wanted to remember the entire experience, while her captors wanted her to forget it. She wanted to interact with them, and ask them questions, while they wanted her to be docile and compliant.

After a while Sandra became aware of a sudden commotion. Another figure had entered the room and was standing behind her. She could sense its presence, and was somehow able to sense its mood. It was displeased, but she was not the source of its annoyance. The little creatures made a sudden move for her, and took off her nightgown. Somehow this amused her; the creature that had entered the room had expected her to be naked, but the small beings had forgotten to remove her clothes, and were apparently being castigated by what she sensed was their leader.

She had a memory of such a creature, which was taller than the others, and somehow more human. Even as she felt herself being disrobed, she felt that she had somehow gained a small victory. Either that, or she was simply reveling in the fact that the creatures who had abducted her were being scolded like naughty children.

Sandra's feeling of elation vanished when she saw one of the short beings holding a tray of decidedly unpleasant-looking medical instruments. There were various small, metallic devices on display. One of them looked rather like a scalpel, and she decided she didn't really want to know what the others were. She knew from experience that some sort of medical operation was about to be performed, and tried to twist her head around to look at the leader being, which she was sure was positioned just behind her. The little beings acted just like robots, or drones, and there was no reasoning with them. *[Many abductees report that the aliens behave like a hive society, with the small beings having very little individuality or independence. Such theories might explain the variations in descriptions of aliens reported by abductees. It may be that they are different types of the same species, just as there are differences in the appearance, behavior and roles of the various type of honeybees within a single nest.]* But the leader being was something entirely different, and Sandra felt that if she could just look at it, she could make it understand.

Her effort to twist her head around was in vain, and the next instant she felt a sharp pain in her stomach. Looking down her body she was horrified to see that one of the small creatures had inserted a long, thin needle into her belly button. *[Such a technique was reported by abductees before it was used by the medical profession in keyhole surgery.]* For a moment she felt nothing, and then she was aware of something moving *inside* her. It was a deeply unpleasant sensation, and as the movement shifted its focus downward, toward her womb, she felt the most excruciating pain she had felt in her life. She screamed in agony and rage, and this seemed to have an effect on the paralysis, diminishing it slightly. She was able to move her head around, and found herself looking straight into the eyes of the leader being. *[Such beings are sometimes referred to as "doctor beings," because of the lead role that they take in the medical procedures that seem to be the focus of so many abductions.]* It was similar to the smaller creatures, but taller; probably around her height of about five and a half feet. Its huge eyes were black, but there seemed to be some movement behind them. She felt instinctively that the being was male, and that she had encountered it before. She yelled at it:

"Leave me alone. I haven't agreed to this."

The being replied instantly: "You *have* agreed. It's all right."

The response was totally unexpected, and she was not even sure whether it had been spoken or communicated telepathically. The content of the response so shocked her that even the pain seemed unimportant. What did he mean? She hadn't agreed, had she? She was assailed by doubt. She knew she had been abducted before, and that there were always parts of the experience that she forgot. Supposing she had consented to this procedure during a previous abduction? But why would she have done such a

thing, and even if she had, didn't she have the right to
change her mind? Equally, the being could be lying. Its response had certainly succeeded in silencing her. *[Con-
spiracy theorists have speculated that this means that
some official agency, purporting to speak for all of humanity, has done a deal with the aliens, and has agreed—
on our behalf—to a program of abductions.]*

The pain was subsiding, but whether this was because
of some anesthetic, or whether the leader being was
somehow diminishing her pain, she didn't know. But as
she looked into its eyes, she felt not hate, but love. The im-
propriety of this response was profoundly shocking to her.
*[The eyes clearly seem to be the focus of control, which
seems to be hypnotic in nature. It is ironic that hypnosis is
used by the aliens to try to induce abductees to forget their
experiences, and by researchers to try and uncover the experiences afterward.]*

The pain had now ceased altogether, and Sandra knew
that the procedure was over. The beings left the room, and
although groggy, she felt sensation returning to her limbs.
She swung her legs off the table, and stood up. The floor
felt cold under her feet, and a shiver passed through her
body. She put on her nightdress, which had been discarded
at the end of the table. Now that she was alone in the room,
a sudden thought occurred to her. Maybe she could steal
something from the craft. If she could take some artifact
that scientists could prove involved an unknown technology, she could prove the existence of the extraterrestrials. But the tools had been removed, and apart from the
table itself, the room was featureless. *[Such attempts are*

frequently made, but even if something is pocketed, the loss is invariably discovered before the abductee is returned.]

She walked out of the room, and found herself in a curved corridor. After a few yards she saw a feature ahead of her, on the left hand side of the corridor, and as she reached it, she stopped in her tracks. Mounted into the wall was a huge tank, filled with a fluid of some sort. The tank was divided into approximately fifty individual sections, and in each one was what looked like an embryonic version of the creatures who had abducted her. A little later on she came to a door, on the right. She went in, and found herself in another featureless room. She was about to leave, when she was aware of a presence behind her. As she turned, she saw the leader being enter the room, leading a tiny creature with him. She sensed that the leader being was pleased, and wondered why. Then she looked at the other being; it was unlike any of the others, and seemed to be more human in appearance.

"She's yours."

The statement had come from the leader being, and she sensed the truth in it immediately. Tears filled her eyes as she looked at the little creature. It was clearly a hybrid, and it seemed very weak. She somehow knew that the medical procedures she had undergone during this and previous abductions were aimed at removing eggs from her. Perhaps this explained the fact that her periods were irregular. She had even had what her doctors had called a phantom pregnancy, although she was convinced that she really had been pregnant. *[So-called Missing Fetus Syndrome, like the debate over implants, is an extremely controversial aspect of the abduction phenomenon. Many researchers will tell you that there is good evidence, but as is the case with implants, it becomes tricky when you try to pin such evidence down. There are many medical reasons why a pregnancy test might prove positive without the person actually being pregnant, and not all miscarriages will contain a fetus, if one never fully develops in the first place. There is, as yet, no convincing evidence for Missing Fetus Syndrome.]*

Her eggs had been taken from her, and fertilized by alien sperm. A hybrid had been created, and possibly grown in those tanks she had passed. And now she knew what was required of her: she had to bond with her hybrid daughter. She held the creature close to her, feeling a deep sense of love. It looked vaguely familiar, like something she had seen in a dream. *[Many female abductees report having what are known as Wise Baby Dreams, where a strange-looking and highly intelligent baby talks to them. Might these dreams be disguised memories of bonding with hybrid babies?]* She was aware of the warmth of her body, and almost felt a spark of something pass from her to the tiny hybrid. Perhaps, she thought, it was a spark of humanity: the vitality that this *child* needed to flourish.

After a few minutes, the leader being moved toward her, and she knew it was time to hand the child back. She started to protest, but he looked directly into her eyes, and somehow stilled any argument. She sensed genuine compassion in the creature, and felt very close to him. *[Such feelings of compassion may be induced as part of the control process, but may also be an example of the psychological condition known as Stockholm Syndrome, where a close bond forms between, for example, a terrorist and a hostage.]*

"You will see her again; I promise."

Somehow this ended any discussion, even before it had begun. One of the smaller beings entered the room, and led the little hybrid child away. The leader being turned to Sandra. It spoke—although she was not aware that its small, slit-like mouth had moved—and a conversation of sorts took place:

"The time has come for you to leave. You will not remember this."

"I will remember. I will!"

"Maybe. We shall see."

The exchange surprised her. It was the nearest she had ever come to any sort of conversation with the aliens. Above all, she craved conversation with them, and longed to move from her position as helpless victim to active participant. She had so many questions, and wished that they would interact with her, one intelligence to another. *[Many abductees go to some lengths to communicate with their captors. Such attempts include leaving various books or drawings on the windowsill, facing outward, or taping a written message to their body when going to bed. Such messages usually contain a plea to ask permission before taking the person, rather than a demand to stop the abductions altogether.]*

The next part of the experience was hazy, and she felt herself to be in a dreamlike state. She was aware of a floating sensation, and of a descent rather like being in a fast elevator. The next instant she hit her bed with a thump, and bounced off, onto the floor. *[This seems sloppy, and researchers have begun to notice increasing examples of such behavior, including several mistakes made by abductors. These have included abductees being put back in bed the wrong way around, or even in somebody else's clothes. If such a trend continues, it may take us considerably closer to definitive proof of the phenomenon—such as two abductees being returned to each other's houses—in different countries. Even the most hard-line skeptics would have difficulty explaining this; although they would probably try!]* For some reason Sandra didn't even try to wake Dave, or write down her experience. She just wanted to sleep. But even as she drifted off, she told herself that she would remember; this time she would remember.

Summary

These two fictitious cases may seem incredible, but none of the details have been made up. All the incidents that I incorporated have been reported before, often in many different cases that have come to light quite independently of one another. I have deliberately avoided drawing any conclusions in this chapter, which was aimed solely at familiarizing readers with the *structure* of an abduction. I will be discussing various theories about abductions in the final two chapters.

CHAPTER 7

Abductions and Science

History has seen many great men and women who struggled against incredible odds to advance the knowledge of the human race. Such sought to push out the frontiers of our understanding, and their vision was responsible not just for ideas, but for the development of these ideas into technology. These people are scientists, and it is no exaggeration to say that they were and are a major factor in shaping the world in which we live. The hallmark of the great scientist has often been the ability to look just that little bit further than the rest of us, and such people have sometimes been persecuted for their beliefs, especially where those beliefs flew in the face of church doctrine. Galileo Galilei ran into severe difficulties with the Inquisition in the early seventeenth century, and was forced to retract what were regarded at the time as heresies, especially his theory that the Earth orbited the sun, and not vice versa.

The issue of proof is a critical one. I accept that I cannot provide *hard* evidence to prove that alien abductions occur, but I believe there to be good enough circumstantial evidence to carry the day. I do not wish to offend readers with religious beliefs, but consider whether the existence of God is supported by hard physical evidence? To put it another way, how many photos are there of God? Religious belief is generally founded upon faith, not evidence. And yet the existence of God is assumed by written and unwritten constitutions all around the world. So if the Establishment accepts the existence of God without proof, why does it demand this proof in the case of phenomena such as alien abductions? The abductees do not ridicule the Establishment's corporate belief in God, so why must the Establishment ridicule those who believe in alien abduction?

Scientists, with their reputation for looking at phenomena at the edge of our understanding, might have been expected to be particularly *attracted* by the alien abduction mystery. But this has not happened, and this chapter will look at the uneasy relationship that science has had with abductions, to try and explain why scientists have never given the subject the serious attention it deserves.

As is so often the case with abductions, we cannot ignore the issue of UFOs. Most of the researchers into the abduction phenomenon came into the subject via ufology, and it is a sad fact that scientists and ufologists generally have a very rocky relationship. Part of this, it has to be said, is because so many ufologists consider themselves to *be* scientists. The very word "ufologist" gives the researcher a sort of respectability that scientists rightly believe must be earned. Some ufologists have brought the whole subject into disrepute and ridicule with their ranting conspiratorial nonsense, and their unprofessional approach. Many people start off with preconceived ideas, and then try to twist every fact to fit their particular theory, rather than approaching the subject in an open-minded way, and seeing where the data leads them. Some people's *need* to believe is so powerful that they are unlikely to let the facts get in the way. Although this is not as common as some might suppose, the vociferous minority responsible for this sort of behavior has biased many scientists against both ufology and the "spin-off" subject of alien abduction. Scientists generally view ufologists and abduction researchers as a group of mildly strange individuals holding broadly similar views. Nothing could be further from the truth. Despite the cranks, the vast majority are serious and sober-minded individuals who are methodical in their investigations, and honest enough to point to conventional explanations where they are uncovered.

Ironically, the first scientific research into the abduction mystery was instigated by ufologists themselves. It should be remembered that in the early days of ufology, the contactees and abductees were regarded with suspicion by the more conservative UFO organizations like NICAP. Such groups felt that these more exotic stories would undermine their research, and trivialize the subject.

Against this background and in view of emerging doubts about regression hypnosis, California ufologist Alvin Lawson decided to set up an experiment designed to test the validity of regression hypnosis as a technique. Lawson carried out his experiments in 1977, together with William McCall, a doctor who had experience with hypnosis. The aim was to see whether people with little or no knowledge of UFOs or abductions could, under hypnosis, imagine convincing accounts of alien abductions. These stories would then be compared with supposedly genuine accounts, and if they were similar, it might be concluded that hypnosis was not the absolute key to the truth that some presumed it to be, but a technique open to fraud or innocent confabulation by over-imaginative individuals. In fact, this is precisely what happened. The "imaginary abductees" produced very intricate stories that bore considerable similarity to the "real" accounts. The subjects had few clues from which to produce such material, being given only a few sketchy details prior to their being hypnotized and asked to imagine an abduction scenario. This is very likely to mirror the situation of most individuals who seek out hypnotherapists to probe UFO and abduction experiences. Most will have a basic idea about abductions, and it is this that will lead them to seek out a hypnotherapist in the first place.

Lawson's experiments should have been welcomed by the UFO lobby, and given prominence within the field. Unfortunately, this did not happen, and believers moved swiftly to condemn his work. Die-hard believers have an unfortunate habit of smearing all skeptics as debunkers, even when the skeptics are within the UFO lobby, and clearly only doing their best to be scientific in their research.

Undoubtedly there were some legitimate concerns over Lawson's work. First, it was argued that the sample of ufological "virgins" may well have had more exposure to the material than they either admitted or consciously recalled. The story of Travis Walton had emerged in 1975, and at the time of Lawson's experiments there was considerable material in the public domain—especially in California, which has always attracted huge numbers of people with colorful beliefs. Another criticism was that Lawson took no account of the lasting emotional trauma displayed by genuine abductees, which was entirely absent from those who had imagined their experience.

Whatever the failings of Lawson's experiments, he had at least tried to inject a modicum of scientific methodology into abduction research. His theories on investigating UFOs and abductions were similarly farsighted. He felt that when looking at abduction reports, the key factor in any investigation should be the witness. He believed that the psychological and physiological state of the witness needed to be investigated in depth, and that this was where the answers to the alien abduction mystery lay. These were wise words indeed, and it is a great pity that more investigators have not heeded them. Experience has shown that one cannot base an investigation into an abduction simply on the raw data that emerges from an interview—whether hypnosis is used or not. One has to look at the history of the witness, examining such factors as their upbringing, their belief system, their physical and mental health, their truthfulness and their previous exposure to abduction material.

A friend of mine with a background in psychology is critical of those ufologists who believe that they alone are qualified to investigate abduction reports. She points out that some abductees may be mentally ill, and that some may be so traumatized by their experiences (whatever their true nature) that they may benefit from professional help. But one rarely comes across instances of abduction researchers steering witnesses toward appropriate mental health professionals or counselors. This is a point worthy of consideration, because if our primary aim is to help the witness, then we are duty bound to ensure that abductees receive the most appropriate help. The Catch-22 here is that because ufologists do not generally have any psychological training, they may not know *when* it might be appropriate to refer someone. Another difficulty is that referral would mean ufologists giving up potentially exciting cases, and the propensity for egomania within certain sections of the UFO lobby means that this may not happen as often as it should. The ideal solution, building upon the ideas of people like Alvin Lawson, would be for ufologists to forge closer working relationships with mental health experts and trained counselors, co-operating on cases, with each respecting the expertise of the other. One organization that has tried to incorporate some of these ideas by recruiting psychologists, doctors and other such professionals is Budd Hopkins' Intruders Foundation, details of which can be found in Appendix 4. In 1992 the organization produced a paper entitled *Suggested Techniques For Hypnosis And Therapy Of Ab-*

ductees. Written by Budd Hopkins and David Jacobs, it attempted to provide some responsible guidance that would be of use to UFO researchers and therapists alike, and was an important step toward building bridges between ufologists and mental health professionals.

Another attempt to bring some scientific methodology to abduction research was an experiment to see whether there were any psychological differences between abductees and non-abductees. Conceived by Budd Hopkins and carried out by Dr. Elizabeth Slater in 1983, it involved nine abductees, and crucially, Slater was not told that the group she was asked to assess had been abducted. Slater administered a full range of standard psychological tests used to measure intelligence and assess personality, such as the Rorschach test and the Wechsler Adult Intelligence Scale—Revised. She then drew up a detailed report on the group, which concluded that they were generally of above average intelligence, had rich inner lives (i.e. they were more likely to daydream or fantasize), and generally seemed to be a little strange. Other characteristics that Slater noted included a tendency toward low self-esteem, lack of emotional maturity and mild paranoia. They were, Slater observed, an extraordinary group. On the face of it, this analysis was not good news for those who believed that abductions happened in a literal and physical sense. It seemed that there might indeed be psychological traits in the abductees that might make it more likely that they had made up their experiences, or misinterpreted something more prosaic.

Budd Hopkins then explained to Slater that the group consisted of people who claimed to have been abducted by

aliens, and asked her to write a supplementary report
taking this information into account. Slater's second
analysis provided Budd Hopkins and fellow believers
with some comfort. She concluded that there was no way
in which the abduction accounts could stem from psychopathology, on the basis that none of the group displayed
any signs of mental illness such as paranoid schizophrenia
or multiple personality disorder. Furthermore, she pointed
out that the peculiar symptoms that she had observed in
the group could well stem from the trauma that would undoubtedly be experienced if the witnesses believed they
had been subjected to the intrusive procedures they described. The argument was, of course, circular. Skeptics
pointed to the mild paranoia and rich fantasy lives of the
group, while believers said these were a natural consequence of the abductions themselves. Whatever the failings of this relatively small-scale research effort, it is
precisely the sort of scientific study that needs to be encouraged if we are to ever understand the true nature of the
alien abduction phenomenon.

A fundamental part of any scientific analysis is measurement. The problem with abductions is that it is extremely difficult to evaluate the scope of the phenomenon.
Although there are several thousand cases that have come
to light all around the world and have been subjected to
detailed scrutiny, this is likely to be the tip of the iceberg.
Many abductees will not know to whom to report their experiences, or may not want to come forward at all, owing
to fear of ridicule, or a desire to forget about their encounters in order to avoid the trauma of dealing with them.
There is also the point that many abductees may not be

aware of their experiences at all. Recalling the case of Steven Kilburn that was detailed in chapter four, it should be borne in mind that there may be vast numbers of people who have experienced elements of the abduction phenomenon but are left with no definite memories—only perhaps occasional feelings of unease, or the odd strange dream or flashback. Many may not even be left with these small clues. Clearly it was vital to try and come up with some idea of the actual numbers of abductions that might be taking place. The most famous attempt to produce such an estimate was the Roper Poll on Unusual Personal Experiences, which was carried out in 1991, and applied the established techniques of market research to assess the true extent of the abduction phenomenon.

The poll had been the idea of American millionaire Robert Bigelow, who has financed much of the more scientific research carried out into abductions. Bigelow had commented to Budd Hopkins and David Jacobs that someone needed to assess just how widespread the phenomenon might be, and market research carried out by an established commercial organization that had nothing to do with ufology was judged to be the best way forward.

It was felt that the survey could not simply ask whether someone had been abducted, because the Roper organization conducted its research by doing face to face interviews, and Hopkins and Jacobs felt that in their experience, people were unlikely to make such an admission to strangers. Furthermore, they believed that many abduction experiences might not be consciously recalled at all. Mindful of these concerns, a series of eleven more general questions were drawn up, five of which were designed to be key indicator questions, where a positive response might suggest that an abduction had taken place. The five indicator questions drew on scenarios from the abduction cases that were known about by researchers, and picked up common themes. The five scenarios were as follows:

Waking up paralyzed with a sense of a strange person or presence or something else in your room.

Feeling that you were actually flying through the air although you didn't know why or how.

Experiencing a period of time of an hour or more in which you were apparently lost, but you could not remember why, or where you had been.

Seeing unusual lights or balls of light in a room without knowing what was causing them, or where they came from.

Finding puzzling scars on your body and neither you nor anyone else remembering how you received them or where you got them.

In addition to these questions there was a check question designed to identify people who might blindly say yes to any questions. This check question asked whether the word "Trondant" had any secret meaning to the respondent. The word had been made up, and anyone who responded positively to this question had their response discarded from the overall analysis.

* * *

Of the 5,947 people quizzed, 18 responded positively to the five key questions and negatively to the check question. This works out at marginally over 0.3% of the sample, and extrapolating this to produce an estimate of how this would apply to the American population as a whole, this would imply that 560,000 people might have made similar responses. These calculations used a total population figure of 185,000,000, because children, prisoners and various other categories had been excluded from the survey. 560,000 potential abductees was sensational enough, but it was then decided to relax the criteria, and include in an estimate of potential abductees those who had responded positively to four out of the five key questions (still discounting those who had responded positively to the check question). One hundred and nineteen people were in this category, and this works out at almost exactly two percent of the overall sample. If this percentage was indicative of the overall population, with two percent of Americans involved, then 3.7 million people might be abductees.

As soon as these results were published, they sparked massive controversy. Critics drew attention to what they saw as two main flaws in the whole project. First, it was pointed out that with the number of positive respondents to all five indicator questions being as small as eighteen, it was meaningless to try to extrapolate this across an entire population. This is a problem with all market research, and it should be pointed out that such criticisms are always leveled against such surveys by politicians—except when the research supports their point of view! The most fundamental criticism was the obvious fact that a positive response to the questions was *not* necessarily indicative that an abduction had taken place at all. All it meant was that a number of people had reported experiences that were also being commonly reported by abductees. There might well be prosaic explanations to all of the scenarios involved.

While these criticisms are not without merit, it should be borne in mind that it was never claimed that the poll would indicate how many of the population *were* abductees; it was merely an exercise in trying to find out how many people *might* be. It is also worth noting that if you look at the percentages of those who responded positively to individual key questions, there are some significantly higher figures. Eighteen percent reported waking up paralyzed and aware of a strange presence, thirteen percent had experienced missing time, ten percent had experienced an unexplained flying sensation, and eight percent respectively had reported seeing balls of light in their rooms, and having anomalous scars. Interestingly, in the responses to two other scenarios that were not treated as key questions, seven percent of respondents had seen a UFO, while eleven percent reported having seen a ghost. Whatever its failings, the Roper Poll is, to date, the best and most scientific attempt to quantify the abduction phenomenon.

When talking about the attitude of scientists to this subject, it should not be forgotten that some of those ufologists who have become involved in abduction research *are* scientists, and *were* scientists long before they were ufologists. Dr. J. Allen Hynek was an astronomer, long before he was involved first with Project Blue Book and subsequently with private research into UFOs and abductions. Similarly, Jacques Vallée is an astrophysicist with a Ph.D. in computer science, while the skeptical Philip J. Klass is an electrical engineer—activities that predate their research into UFOs and abductions. The British psychologist Dr. Susan Blackmore has done some interesting research into abductions, while even the great psychiatrist Carl Jung formulated some theories about UFOs that might also be applied to alien abductions. We shall be looking at some of these theories in chapter fifteen. But when considering scientists who have taken an active interest in the abduction mystery, and made a major contribution to the field, one is drawn inexorably to the work of Dr. John Mack, Professor of Psychiatry at Harvard Medical School.

John Mack's involvement in the subject dates back to a meeting with Budd Hopkins in January 1990. Previously, when asked if he was interested in meeting Budd, and hearing about his research, Mack had declined, on the grounds that the whole business sounded dangerously odd. Mack's eventual meeting with Budd Hopkins convinced him that there was enough to the subject to justify serious investigation, which he began in the spring of 1990 when he carried out his first interviews with abductees. As a professional and highly respected psychiatrist with over forty years of clinical experience, skeptics might have expected John Mack to conclude that there were internal, psychological causes underlying reports of alien abductions. This was not to be. Mack found no evidence of any psychiatric disorder in the abductees (aside from trauma resulting from the experiences themselves), and concluded that whatever the true origin of the beings responsible for the abductions, they *did* manifest in the physical universe. Mack's book, *Abduction: Human Encounters With Aliens*, caused a sensation when it was published in 1994. This was hardly surprising, not least because Mack's investigations had uncovered material that even some of the believers found difficult to accept at face value. His investigations (using therapeutic interviewing together with hypnosis modified by Grof breathwork techniques) uncovered a number of instances of dual human/alien identity, and various past life memories, some of which involved the abductee having lived before as an alien.

Mack's approach was unconventional, and he felt that Western scientific methodology was insufficient to explain the abduction phenomenon. He favored more of an Eastern approach and believed that a more spiritual emphasis was needed, arguing strongly for what he termed a "paradigm shift."

One of Mack's most controversial statements concerned child abuse. It had been suggested that some "memories" of abductions might in fact be distorted memories of childhood sexual abuse. Mack not only maintained that this was unsubstantiated by any single case he was aware of, but went on to say that on several occasions investigation of someone who believed they had been the victim of childhood sexual abuse had turned up evidence of an alien abduction.

All this proved too much for the authorities at Harvard, especially when Mack appeared on a number of

American television shows to talk about his findings. The Board at Harvard Medical School launched a formal inquiry into his conduct, setting up a committee that was chaired by the medical school's emeritus professor, Dr. Arnold Relman. Mack's critics said that his research had been sloppy and claimed that his dealings with abductees were inappropriate. They felt that he had brought Harvard into disrepute, and argued that he should have submitted his findings for peer review in the academic press, rather than proceeding in the very public manner that he did. Mack countered by saying that he had tried to publish in an academic journal, but had his paper rejected. With regard to peer review by his colleagues, Mack had invited several psychiatrists to sit in on some of his sessions with abductees. Several of these colleagues offered testimony in support of Mack, as did others who regarded the inquiry as an unwelcome challenge to the principle of academic freedom.

After exhaustive meetings, the committee concluded that Mack had the right to investigate any subject that drew his attention, but criticized his methodology. The committee had no authority to evaluate Mack's claims, and pointedly avoided offering any view on the reality or otherwise of the alien abduction phenomenon. The report was forwarded to the medical school's Dean, Daniel Tosteson, who publicly censured Mack, but told him that his job—which had been under threat—was secure.

The inquiry had split the academic community, and one senses that the real objection behind the criticisms leveled lay not with Mack's methodology, but with the material itself. It was simply *too* bizarre for the often conservative world of academia to accept. The episode is significant because it shows how the world of science can be divided when it studies phenomena at the very limit of human experience. Mack had launched a two-pronged assault on the academic establishment. First, his research had taken him into an area that many scientists regarded as nonsense. But secondly, and more fundamentally, he had dared to speak the unspeakable by claiming that in certain areas a conventional scientific methodology (with the emphasis on measurement and repeatability) was simply not adequate to the task. He felt that the abduction phenomenon needed a new scientific paradigm. This, of course, is nothing less than a challenge to science itself, and an attempt to redefine fundamental concepts of proof and reality.

Professor David Pritchard is a physicist at the prestigious Massachusetts Institute of Technology (MIT), who specializes in atomic and molecular physics. He is also one of the growing number of respected scientists who have come to regard alien abductions as being worthy of serious, scientific study. But as much as the subject fascinates him as a scientist, he abhors the way in which abduction research attracts so many cranks and charlatans. He had considered writing a book, before he decided that what was really needed was a conference. The conference he had in mind would not be the usual sort of conference organized by ufologists, but a gathering of scientists, all trying to bring their own specialist expertise to bear on the subject. It was a tall order, but from June 13–17, 1992 the world's first scientific conference on alien abductions took place at MIT, co-chaired by David Pritchard and John Mack. Much of the funding was provided by Robert

Bigelow, who had previously funded the Roper Poll on Unusual Personal Experiences.

The conference heard from researchers such as Budd Hopkins and David Jacobs, and from some of the witnesses themselves. But a multitude of scientists from various different disciplines were present, including doctors, psychiatrists, psychologists, sociologists, hypnotherapists, physicists and astronomers. Such a disparate group was unlikely to reach a consensus on what lay behind the abduction phenomenon, but the conference's main achievement was not so much in providing solutions, but in showing that despite their reputation for conservatism, some scientists were making attempts to investigate alien abductions. The proceedings of the conference were written up and published under the title *Alien Discussions*. This heavyweight tome—details of which can be found in the bibliography—is essential reading for all those with a serious, academic interest in the alien abduction phenomenon.

There is clearly much work to be done, and many people yet to be convinced. Although he was a driving force behind the various Search for Extraterrestrial Intelligence (SETI) projects carried out by both NASA and others (using radio astronomy to look for signals from extraterrestrial intelligences), the late Dr. Carl Sagan was highly skeptical about alien abductions. In his last book, *The Demon-Haunted World*, Sagan launched a blistering attack on what he categorized as unscientific research into UFOs, alien abductions, and a host of other mysteries.

Ironically, such skepticism comes against a background of the most sensational advances in scientific understanding—advances that serve only to make the existence of extraterrestrial visitors seem more likely. In September 1995 the journal of the Royal Astronomical Society published a paper by Dr. Ian Crawford, entitled *Some Thoughts on the Implications of Faster-Than-Light Interstellar Space Travel*. Crawford is an astronomer at University College, London, and his paper (which had to pass the traditionally strict independent review process that precedes publication in any scientific journal) theorized that true interstellar travel might be possible within the existing laws of physics. This would utilize the concept of wormholes in space-time—areas where different laws might apply. For anyone who thinks this sounds like science fiction, pick up one of any number of books on cosmology or quantum physics, and you will see that scientists are developing theories about the nature of the universe that make the alien abduction phenomenon look positively mundane!

Meanwhile, other astronomers had proved the existence of planets around other stars, not through visual observation, but by studying the way in which they affected the stars' orbits. Until then, there was no physical evidence to refute the theory that ours might be the only solar system in existence anywhere in the universe, however unlikely this seemed.

On August 6, 1996, NASA made an announcement that changed the world of science forever. Fossilized traces of what seemed to be primitive bacteria were found within a meteorite that had originated on Mars. Although still the subject of some controversy and debate, the announcement seemed to be nothing less than *official* confirmation of the existence (at least at some stage in the past) of extraterrestrial life.

Such quantum leaps in our understanding of the universe and our place therein seem to me to provide the foundations upon which scientific acceptance of at least the *possibility* of alien abductions might be built. But conservative, Western science sets very definite criteria when it comes to the concept of proof. It is a logic-based methodology with the emphasis on observation and repeatability. But these are *our* rules, and the visitors—whoever or whatever they are—play by *theirs*.

CHAPTER 8

Defense of the Realm

Life can be a dangerous business. Domestically, people face threats from crime, and for this reason we have a criminal justice system, consisting of various elements such as the police, the judiciary and the prison system. Another domestic threat is the ever present fear of illness or injury, and for that purpose we have a multifaceted healthcare industry, consisting of elements such as the National Health Service and medical insurance. The other obvious threat is that of aggression from other countries, which in the worst-case scenario could lead to war. It is to deter such a threat, but also to deal with it should it materialize, that we have the Armed Forces, and the Ministry of Defence. The point of these introductory remarks is simply to illustrate a self-evident fact of life. We all face a series of threats in life, but where something can be done to eliminate, reduce or even prepare for such threats, we expect the State to take action where it can. Of course, *we* all have responsibilities too, and people are generally expected to be individually accountable for their own actions, but at a national level the primary duty of any democratic government must be to protect its citizens from harm.

We live in a very uncertain world. Sometimes the old threats diminish or disappear, as with the way in which the Cold War ended. But sometimes new threats emerge to torment and challenge nations, and when this happens (as with diseases such as mad cow disease) we expect our governments to respond quickly, in order to safeguard the public.

Is alien abduction not such a threat? Whether or not one believes these events are extraterrestrial is irrelevant, because the fact is that people *are* suffering as a direct result of the phenomenon. I, and many other researchers all around the world, have cases involving *something* that intrudes, uninvited, into people's lives. Should governments not be doing something about this situation?

When, during my official duties, I was first asked about alien abductions, I was surprised to find that the Ministry of Defence had no official policy on the alien abduction phenomenon. Technically, this meant that any research or investigation was outside my authority. But such a "not in my job description" mentality has never been the way in which I operate, and it was simply not an option when abductees contacted me, often in a distressed state. If they wanted to talk to a government representative, there was no one else to refer them to in any case. I always found time to talk to abductees, and at the time I did my best to help them come to terms with their experiences, empowering them, and trying to help them to help themselves. This was a difficult business and I had to tread very carefully. I am not a mental health professional or a trained counselor and, although I do not believe such individuals are necessarily any more (or less) able to evaluate the phenomenon than ufologists, they are experienced at spotting signs of psychopathology that others might miss. I decided that the most important skill to develop was the ability to listen. In a situation where there are so many unknowns, and few right or wrong answers, I tried to respond with empathy. I found that I was often the first person that someone had confided in, and if you find this hard to believe, ask yourself how your loved ones would react if you told them you believed you had been abducted by aliens. I felt honored and privileged that people were sharing secrets with me—a stranger—that they had not shared with their partners, families or friends. In fact, this sharing of personal information with strangers is often the key to talking about things that an individual has difficulty in facing, and is something that lies at the center of much psychiatry and counseling.

With this sharing of information came responsibility. For example, I was often asked whether I recommended regression hypnosis as a way of unlocking hidden memories. Although for reasons that I have already explained, I am generally against this technique, this was not my choice to make. I tried to point out the advantages and disadvantages, but said that ultimately this was a decision that only the individual could make.

My primary aim within the limited resources available to me was not so much to investigate the cases, but rather to help the abductees come to terms with their experiences. I also believed it was important to avoid any implication that cases could necessarily be "solved" in the same way as a conventional crime. One case is unlikely to unlock the entire mystery. Similarly, with so many possible explanations of what was going on, I felt it was irresponsible to promote any individual theory about what abductions really were. So I found myself listing a range of options, and explaining that there were a number of possible interpretations to each individual experience.

I justified my official involvement by reference to the point mentioned previously—namely that governments had a duty to protect their citizens from *any* threat, *whatever* its nature or source. Or as one irate abductee put it to me, in a rather more direct manner:

"You work for the Ministry of Defence . . . so bloody well defend me!"

Throughout my official work with abductees I was having to justify my actions to one or two closed-minded individuals who were trying to dissuade me from getting involved at all. What I desperately needed was some official authorization that could have given me the authority to continue my involvement without constantly having to justify my actions. But the Ministry was never going to officially sanction research into alien abductions, so some other way had to be found. As it happens, in my opinion such a way does now exist, and official work on the abduction mystery could now be fully justified by reference to one of the Ministry of Defence's *own* key documents.

The Ministry of Defence sets out all its responsibilities in an annual report that is presented to Parliament by the Secretary of State of Defence. *Statement on the Defence Estimates 1996* describes Defence Role One as follows:

> *To ensure the protection and security of the United Kingdom and our Dependent Territories even when there is no major external threat.*

Defence Role Two reads:

> *To ensure against a major external threat to the United Kingdom and our Allies.*

These general Defence Roles are subdivided into a series of Military Tasks, and Military Task 1.10 is of particular relevance. It is entitled "Maintenance of the Integrity of British Airspace in Peacetime" and reads as follows:

> *The integrity of British airspace in peacetime is maintained through a continuous Recognised Air Picture and air policing of the United Kingdom Air Defence Region.*

Such words were designed, of course, to cover the threat of aggression from foreign powers, and to cover other scenarios such as terrorism. But some ufologists have pointed out that the document is open-ended enough in its general talk of ensuring against threats to sweep up the concept of alien abduction. Nowhere in those Defence Roles does it stipulate *where* the threat must come from.

The situation is easier to look at in relation to incursions into controlled airspace by UFOs. Over the years there have been numerous incidents where RAF aircraft have been scrambled to try to intercept uncorrelated targets picked up on military radar systems. Sometimes these turn out to be something conventional, such as Russian maritime patrol aircraft probing at our air defenses. But on other occasions there are no conventional explanations.

Now consider again the words of these Defence Roles and Military Tasks. It could be argued that they convey upon the Ministry of Defence a duty to protect the citizens of the United Kingdom against *any* threat posed as a result of unauthorized penetrations into British airspace.

If the official response to breaches of our air defenses by UFOs is perceived by some as inadequate, in relation to alien abductions, it is nonexistent.

This was illustrated by a letter sent to the Ministry of Defence by an abductee, asking for an official view on the alien abduction mystery. I sought and obtained permission to reproduce this letter, which is at Appendix 1. The response from Secretariat (Air Staff) 2a at the Ministry was, in my view, extraordinary to say the least. After proceeding to set out the Department's standard line on UFOs, the reply went on to state:

> *Abduction is a criminal offence and as such is a matter for the civil police.*

The Ministry's letter is reproduced in full at Appendix 2.

So that was it. In a few words the Ministry had washed its hands of the whole business, and effectively passed the problem to the civil police. What are they supposed to do to prevent people being abducted by aliens, one wonders. Give them a ticket?

Although I felt that the Ministry of Defence was simply trying to avoid getting embroiled in a difficult subject, it did seem worth following up the idea of abduction as a criminal offense, however bizarre it sounded. Accordingly, I did some research, and found out that the advice was not entirely correct. Abduction *is* a criminal offense, but only in the context of the Sexual Offences Act 1956 and the Child Abduction Act 1984. But none of the sections of these statutes are relevant to the *alien* abduction phenomenon, where more applicable offenses include kidnapping, false imprisonment and assault.

Meanwhile, the abductee who had written to the Ministry of Defence had sent a letter to the Sentencing & Offences Unit of the Home Office, which handles policy issues on matters concerning Offences Against the Person. This letter—which is reproduced at Appendix 3—asked whether the Home Office or the police had ever taken any action as a result of the numerous public reports of alien abductions. To date, no reply has been received.

This whole exercise, I felt, mirrored something that had taken place in America in 1987, when Budd Hopkins wrote to the FBI in an attempt to get them to take the abduction phenomenon seriously. He enclosed a copy of *Intruders* and stressed the trauma that had been experienced by many of the abductees that he had interviewed. This action was the culmination of a disagreement between Budd Hopkins and Philip J. Klass, who is notoriously skeptical about abductions, and critical of many abduction researchers. Klass had asked why nobody had reported events to the authorities, and was clearly convinced that an approach to the FBI might expose how little evidence there was for alien abductions. As it happened, little came of the initiative, largely because—as discussed in the previous chapter—the abduction phenomenon is unlikely ever to be validated by reference to an empirical methodology of the sort employed by the FBI.

Notwithstanding the above, there is one area where some interesting data might be lurking, and that is within statistics about missing persons. Governments seem to treat this whole area with disdain, regarding people who are reported missing as social losers who have probably run away from home and drifted toward the anonymity of big cities, where they disappear into a sea of troubled humanity. Clearly many of the people reported missing each year *have* simply run away, and tragically, some have probably been murdered. But might these figures hold a terrible secret? When we study alien abductions, we study cases where the abductees have been returned. Might it be that not all abductees are returned? Is this not something that should be considered as a matter of urgency by governments and abduction researchers alike? And if missing persons statistics might conceal abductees who were never returned, what might the position be in countries where no proper records of missing people are kept? Thousands of people might be going missing, and we would be none the wiser.

The previously mentioned interdepartmental debate about responsibility for alien abductions mirrored an embarrassing incident that occurred just after the 1996 release of the science fiction film *Independence Day*, in which aliens attempt to invade the Earth. The *Evening Standard* newspaper wondered which department would take the lead in the event of overtly hostile action from extraterrestrials, and whether some sort of contingency plan existed. They contacted the Ministry of Defence, who argued that such a situation would constitute "civil defense," and would therefore be a matter for the Home Office! When the Home Office were duly approached they told the reporter not to worry, on the grounds that if an alien invasion did occur, the Ministry of Defence would leap into action!

The abductees themselves regard such squabbling between different government departments with a mixture of amusement, anger and contempt. They are not particularly concerned whether their cases are investigated by the Ministry of Defence or the Home Office, so long as *somebody* within the ranks of officialdom is prepared to come to grips with the subject. My view is that this must be a matter for the Ministry of Defence, because this is the only department that might actually be able to *do* something about the situation, if only it would try. This may sound fanciful, but it is my belief that meaningful action could be taken. If abductions really do take place in the physical sense, the military may already possess the tools that would enable them to validate—and perhaps even prevent—abductions. It is simply a case of coordinating these assets, retargeting them to the specific mission in hand, and looking at the resultant data with an open mind. At the moment the military aren't finding evidence of abductions—because they aren't looking for any!

So what could be done? The first line of defense consists of long range surveillance systems, such as military spy satellites. Satellite imagery of the site of an alleged abduction might pick up any UFO that might have been there at the time. At the moment these assets are targeted elsewhere. It would be comparatively straightforward to use these systems to assess better what is going on in our own country.

The second line of defense is radar. Although some UFOs are entirely stealthy, my contacts with radar operators suggested that UFOs were frequently picked up on radar, but that RAF jets were generally scrambled only if the intruder behaved like a conventional aircraft. But we are not dealing with conventional aircraft, so we are left with a bizarre situation where uncorrelated targets appear on the radar, but our air defense fighters stay on the ground. It would be easy to change the standard operating procedures and give fighter controllers more leeway to scramble aircraft. We might come closer to intercepting craft that, at present, appear to operate in our airspace with impunity.

The final line of defense would be Special Forces units, deployed on covert surveillance of the properties of those who have reported abductions. All research indicates that abductions are not once-only events, but part of

an ongoing process, so perhaps it would be possible to
hunt the hunters. This has been tried before by abduction
researchers, who have used all sorts of video equipment
and commercial security apparatus in their efforts to prove
that abductions take place, and to attempt to stop them.
But such attempts have come to nothing, because the abductees themselves turn the equipment off, as if under
some form of compulsion. I therefore suggest that surveillance needs to be mounted *independently*, without the abductee being aware. This raises serious issues about
privacy and consent, but since the aim would be to prevent
intrusive procedures that themselves occur without consent, perhaps this is a price worth paying. This may sound
somewhat extreme; but if abductions occur on a physical
level, then do we not have a responsibility to investigate
such claims to the best of our ability? There is little to lose
and much to gain by taking the few practical actions suggested here. If nothing else, governments could then at
least claim that they had done their best to investigate the
serious allegations being made by their own citizens.

So if there are things that could be done to prevent abductions, is it possible that such things have been attempted?
Is the Ministry of Defence covering up knowledge of alien
abductions? I do not believe this to be the case. I subscribe
to the view that what we are witnessing is failure to come
to grips with a patently real phenomenon—rather than a
sinister conspiracy. As some researchers have pointed out,
if the Ministry of Defence is covering anything up here, it
is more likely to be ignorance and prejudice rather than
any hard data. It has always been apparent to me that, as
is the case with UFOs, the Ministry knows absolutely

nothing about the abduction phenomenon. I suppose it is a difficult and embarrassing admission that there are things being done to UK citizens that the government is power-less to prevent. If alien abductions occur in a physical sense, then they illustrate the ultimate failure of defense.

The idea of a cover-up on the subject of abductions may sound bizarre, but there have been persistent rumors of such a conspiracy in America. These rumors seem to be based on little more than an extension to the idea of the UFO cover-up. In other words, if the American gov-ernment knows all about UFOs, and if the occupants of UFOs are responsible for abductions, then the United States government must know about—and perhaps even sanction—abductions. The theory is that the US govern-ment has a deal of some sort with the extraterrestrials, and that under the terms of this agreement the aliens are given a free hand to carry out a program of abductions. It is al-leged that in exchange for this the Americans have re-ceived some limited technological assistance from the extraterrestrials.

Some researchers believe that such a deal was struck immediately after the Roswell crash in July 1947, which forced the extraterrestrials to show their hand. The US Army got to keep the craft, it was claimed, but in doing so signed away the democratic rights of the American public. It is alleged that any such deal remains an ongoing con-cern, and that recent benefits to the Americans have in-cluded the development of stealth technology, and the acquisition of a number of intact alien craft, which are supposedly being test flown in Area 51 in Nevada—a place sometimes known as Dreamland.

Before you protest that no government would ever make a deal of this sort and treat its citizens in such a way, remember that in the dark days of the Cold War an offer of quantum leap technology would clearly have been very tempting. But talk of a deal implies some sort of equal partnership. This is unlikely to be the case, and even with the fact that extraterrestrial visitors would have a self-evident technological advantage over us, it may be that we have been forced into such an agreement, rather than having entered into it willingly. Although this is intriguing speculation, my personal view is that it goes too far. The US may *suspect* that abductions occur, but probably has no direct knowledge. The British government, I am sure, hasn't got a clue!

The debate about whether governments should be making efforts to counter—or at the very least investigate—alien abductions tends to concentrate on arguments that are almost legal or constitutional in nature. As such, it is easy to lose sight of the individual players in these events—the abductees themselves. This was always a danger for me at the Ministry of Defence, because there were occasions when I was fighting a desperate rearguard action to justify carrying out even the most cursory of investigations. But all the while, the abductions continued. The cases still came in.

Now it's time to move on to the second part of this book, which starts with a number of cases that I have investigated.

PART 2

THE WITNESSES

CHAPTER 9

Parallels

Many researchers believe that abductions follow a standard format, with a clearly defined structure. They believe that there are, in effect, rules to the game. While I accept there is considerable commonality in some people's experiences, as I outlined in chapter six, we must never forget that there is also great diversity in the phenomenon. It involves individuals, and is interpreted through the eyes of these individuals, reflecting and even shaping their own beliefs.

Patsy (pseudonym) has chosen to remain anonymous because—like many who have experienced unexplained

phenomena in their lives—she does not wish to be publicly labeled, either as an "abductee" or an "experiencer,"
and she rejects the idea that her strange experiences
somehow define who she is. I have omitted specific details
of the location of some of these events, to further protect
her identity. Patsy is an intelligent London-based woman
in her late twenties, well-educated and articulate. Patsy's
story is different in many respects from most cases I have
investigated. It is worthy of inclusion nonetheless, if only
to remind us that abduction research is never easy, not
least because we do not know where the boundaries of the
phenomenon lie.

Although Patsy has experienced many peculiar things
in her life, she herself doubts that she is that different from
anybody else in this respect. It is, perhaps, simply that she
is analytical by nature, and open to the idea that reality is
not the clearly defined concept that many believe it to be.
Patsy's experiences are not the ghosts and UFOs that most
people might associate with the concept of unusual phenomena, but something altogether more exotic. They include precognitive areas, synchronicity and an awareness
of those closest to her that can really only be described as
psychic.

One of Patsy's first memories of something out of the
ordinary is a recollection of encountering a huge black
crab in the back garden of her parents' house, when she
was a young child of about four. She clearly recalls that it
was about a foot long, and remembers how she shouted for
her mother, who killed it with a shovel, placing the dead
creature in the garbage can. At the time, Patsy knew instinctively that there was something strange about the

event. She tried to rationalize what she had seen, and wondered whether the crab might have escaped from a nearby fish and chip shop. Looking back on the incident as an adult, Patsy convinced herself that the idea of a huge crab in the garden was so unlikely, that the creature must have been some sort of gigantic spider. This was despite the fact that her recollection of the crab was absolutely solid. To this day, Patsy suffers from arachnophobia.

This incident might suggest a screen memory, where the mind constructs a false memory to hide a traumatic or inexplicable truth. If what is being seen is totally beyond the experience of the witness, then the mind simply makes the best approximation it can, drawing upon available memories. This is the same process whereby less technologically advanced societies that have encountered an airplane do not see a "metal bird," but simply a bird; it is the closest match. Some researchers might claim that the creature Patsy saw was actually a small, dark alien. There are certainly cases where abductees have encountered owls, deer or cats that subsequent investigation has suggested might not have been all they seemed. The clue here is the eyes, however, and the eyes of the crab (or spider) are not something that Patsy especially recalls. Some ufologists fall into the trap of trying to link *any* anomalous experience to an abduction. This is plainly nonsense; alien abductions are only one of many strange things going on in the universe that are not yet fully understood. Exactly what it was that the four-year-old Patsy encountered, we shall probably never know and Patsy has no desire to raise the issue with her mother.

Patsy's next experience occurred when she was about

seven years old, and visiting Italy with her family. The family were in the elevator of a large department store when all of a sudden it stopped between floors. After a few moments, a palpable feeling of panic was discernible, and at this point Patsy's memory suggests that something unusual had happened. She recalls being at the front of the elevator, and prizing open the doors. She stepped out into what she could only describe as a circular landing of some sort, being perhaps twenty-five or thirty feet in diameter. The "landing" was white or light gray in color, and featureless apart from a series of double lift doors, visible all around the edges of this bizarre room. There were perhaps twenty such elevators in total. Patsy noticed a spiral staircase on the left hand side of the room, and started to climb it. But when she got about halfway up, she saw that the next floor was identical to the one she had just come from, and she decided to head back to the elevator, because she felt that there was no other way out.

Despite her strange experiences, Patsy is often extremely skeptical, and always strives to rationalize them. On one level she feels that this incident might have been some sort of escape fantasy, designed to avoid the trauma of being trapped in an elevator. The alternative, however, is that the whole event occurred in the most literal sense, although it is difficult to believe that a seven-year-old child would take the lead in forcing the elevator doors, escape into a singularly unlikely room, only to return to a faulty elevator to find her family and the rest of those trapped there still in it. This "it couldn't have happened therefore it didn't" attitude is, however, a typical adult denial. At the time Patsy was certain that the event did indeed take place in a literal and physical sense, and she still remembers it as such.

Some ufologists might suggest that Patsy was abducted while in the department store, and that many of the memories that she has are distorted recollections of the interior of a spacecraft, in the familiar circular shape. Curiously, I have investigated more than one abduction case where the person concerned has a profound fear of elevators, and it is entirely possible that this may have something to do with a negative reaction to the central memory of being taken *up*, into a craft.

That Patsy could have been abducted from a crowded elevator might seem bizarre, and even those who accept the reality of abductions might find it hard to believe that such events occur anywhere else apart from in bedrooms and on lonely roads at night. Yet there *are* instances of abductions occurring in more public places. Budd Hopkins has one celebrated case involving a woman who was apparently abducted from a cocktail party!

Putting aside abduction theories, it is possible that there is an altogether different explanation. Patsy is a firm believer in parallel universes and other realities—beliefs that are now backed not just by science fiction but by quantum physics. Might Patsy not somehow have encountered a junction between realities?

While the two events mentioned so far are undoubtedly strange, I would not have included Patsy's story in this book had it not been for an event that occurred in the summer of 1992. It is the sort of experience that sends abduction researchers into a frenzy, especially since this event—like the childhood events described previously—was *consciously* recalled. Patsy has not been regressed, and indeed has never been hypnotized in her life.

It was a hot day, late in June, and Patsy was relaxing in a London park, where she had gone to spend a quiet afternoon reading and sunbathing. Lying on her back with her eyes shut, she suddenly felt a strange spinning sensation, and to her utter amazement felt herself go spiraling up into the air, as if she were being sucked into a vortex. The sensation was a violent and unpleasant one, but her reaction to it was even more extraordinary. Instead of being afraid, not least because she seemed to be about fifty feet above the ground, she was *embarrassed*. She was worried that other people would see her up in the air, and wonder what she was doing up there. Such an inappropriate response might normally be indicative of a dream, or of some altered state of consciousness. There is a possibility that the event was an Out of Body Experience, or OBE, which some people believe involves the awareness, spirit or soul somehow leaving the physical body. But as we shall see, there was to be a dramatic and compelling piece of evidence that would support the theory that a *physical* event of some sort had taken place.

Patsy's immediate concern was to get back down to the ground. She concentrated hard, and willed herself to return. She came down slowly but surely, although she is clear in her mind that the last part of her descent was the hardest. While she has no recollection of seeing her own body on the ground from her airborne position, she remembers a strange meshing sensation, which she believes was her spiritual body interlocking with her physical self. The final coupling was marked by a singular jolting sensation, as if the two separate selves had clicked together.

The physical evidence that was to become so *vital* was preceded by a pain. Patsy is unsure whether the pain started *before* the sensation of leaving her physical body, or afterward. This confusion is interesting in itself. In my own investigations I am always wary of people who seem to have all the answers, since in my experience, the most plausible cases are often those where the witness is unsure of some of the detail, and punctuates his or her testimony with phrases such as "I don't know."

Patsy is a fairly tough individual, and has endured some horrendously painful backaches in her life. But she is insistent that the pain on this occasion was the most excruciating that she had ever experienced. It was a highly localized sensation, concentrated in her upper thigh. She wanted to writhe around on the ground in agony, but felt that she couldn't do so in a public place. She wanted to look at the cause of the pain, but was unable to do so without taking off her shorts. The pain became increasingly pronounced, until she had to leave the park. Oddly, after making this decision she has no recollection of looking at her leg, which would logically have been the *first* thing she would have done once she arrived home. I quizzed Patsy vigorously about the events that followed her leaving the park, but she was insistent on this point. "Do you want to know what I remember? Nothing nothing nothing!"

When Patsy did finally look at her thigh, she was horrified. There was a mark there, in the exact spot where the pain had been. The wound was about two centimeters long, and one and a half centimeters wide. As time went on it grew more pronounced, becoming redder and more raised. She showed it to her friend Hazel (pseudonym), a nurse with ten years' experience, who said it looked like a burn.

Two or three days later, the pain, which had apparently subsided over the intervening period, began rapidly increasing in intensity, and had spread to the rest of her leg. Going out with two friends to celebrate a birthday, she had to take a walking stick with her, as she was unable to walk unaided. At some stage in the evening she showed the injury to her friends, who were shocked at its alarming appearance. A little while later the pain became so intense that Patsy had to cut the evening short, and ask Hazel to take her to the emergency room of Guy's Hospital. Being a nurse, Hazel was well aware of the demands that were placed upon emergency staff, and would not have wasted their time on anything trivial. Her insistence that Patsy go to the hospital was testament to the seriousness of the situation.

By this time Hazel had extracted from Patsy the details of her strange experience in the park. As it happened, Hazel was aware of the abduction phenomenon, and raised this possibility with Patsy. Patsy gave her a withering look; it was not the first time that she had been sounded out about alien abduction, and on a previous occasion she had described someone who believed that they had been abducted as "a flake."

At the emergency room the nurses examined the wound and confirmed Hazel's initial view that it looked like a burn. But Patsy was adamant that this simply could not be the case. She had no recollection of having received such an injury, and would surely have remembered one if it had occurred. Patsy was more concerned that she might somehow have been bitten by a poisonous spider, and asked the nurse to check this possibility. There is an interesting parallel between this thought and the way in which Patsy's "rationalization" of her childhood sighting of the huge crab also involved a spider. The nurse who had given the original opinion sought the view of a doctor, and he too confirmed that the wound looked like a burn.

The wound slowly healed, but the scar remained, and is faintly visible to this day. It is reminiscent of the sort of scars often reported by abductees as stemming from a medical operation carried out by extraterrestrial beings. Whatever the facts of this case, the questions still remain: how did Patsy receive such a serious burn in a public place? How was it that she was not aware of how she received the injury? How did it occur without leaving any mark on her shorts?

None of the experiences described in this chapter so far are obviously indicative of an alien abduction, and at first glance such an interpretation may seem fanciful. There are, after all, no UFOs to be seen, and no aliens. But there are parallels. The final experience mentioned here is similar. It occurred during the afternoon of Saturday September 23, 1995 while Patsy was working in a central London art shop. Saturdays were usually the quietest days, because the shop was away from any main roads, and there was no passing trade from the office staff who frequented the area during the week. All of a sudden, a group of people assembled outside the shop, and Patsy commented to one of her colleagues, Maggie (pseud-onym), that it looked as if a bus tour had arrived. The group began to filter into the shop, and both Patsy and Maggie felt distinctly uneasy. Patsy remembers laughing nervously. Something was wrong, but neither of them could quite put their finger on it.

The people who were filling the shop were definitely unusual. All of them were oddly dressed, as if they were uncomfortable in their clothes. One middle-aged man, for example, was wearing a stylish pinstriped suit, but had a pair of scruffy running shoes on his feet. Another was wearing an open-necked shirt, a waistcoat and a ban-danna, in the style of a cowboy. They were a mismatched group, who did not look as if they belonged together. There seemed to be many different styles of clothes, and many different accents, which is not what one would have expected if they had been a bus tour. A man with what ap-peared to be an American accent said hello to Patsy, but she found herself unable to speak at all, and simply mouthed a response. Maggie said to Patsy, "Oh my God, they must be aliens."

It was a most peculiar thing for Maggie to say, be-cause she had no knowledge or interest whatsoever in UFOs or abductions. For some reason Patsy decided to count the people, and it transpired that there were thirteen of them in the shop together. The shop is a specialist one, and thirteen people was the most that Patsy had ever seen in the shop at the same time. Strangely, for the duration of their visit there were no other customers present. It was as if an otherworldly atmosphere had descended, rather akin to the Oz Factor that has been described elsewhere. All of the visitors were talking, but even though Patsy and Maggie were only two feet away from the nearest of them, they were unable to make out a single word. It was as if there was general background noise, but no distinct speech. Another individual approached Patsy, and although she was unable to make out any distinct words, she thought that there was a hint of a northern accent. Patsy was still unable to speak, and all she could do was to mouth a greeting.

The telephone rang, and Patsy clearly heard Maggie telling someone: "The shop's really busy; it's full of aliens."

The group finally began to filter out of the shop, but at this point a Swedish-looking woman stepped forward and made a purchase of a single pencil—the only thing that the entire group bought. There was an awkward moment during the transaction when Maggie handed the woman her receipt. She looked at it strangely and then handed it back. She appeared confused, and the clear impression was that she did not grasp the *concept* of a receipt.

Patsy went downstairs for a fifteen-minute tea break, and was coming back up the stairs, when a sudden thought came to her. She has great difficulty in describing the process. It was not exactly telepathy, but it was as if something profound had just *appeared* in her awareness. The idea that struck her was that the entire situation that she had witnessed was designed to teach her about dealing with groups of people. When she got to the top of the stairs, the shop was indeed full of people, and again, this was unusual for a Saturday. But now things seemed to be happening as normal; conversations could be heard, and purchases were being made. Maggie, meanwhile, had gone for her break, and Patsy asked the third member of staff if Maggie had told her about "the aliens." Maggie hadn't, which was strange, given that it had been a quiet day and something highly unusual had just occurred, which Maggie had previously been discussing at great length. Is it possible that the experience wasn't mentioned because it had already been *forgotten*, in a way that parallels the amnesia of the abductees?

Patsy has flirted briefly with the world of ufology, and decided that it is a world she wants little to do with. As a psychology graduate she is well aware that the complexities of the human mind offer some more prosaic explanations for alien abductions, and as a counselor she is dismissive of the way in which many ufologists seem to exploit the abductees, using them as material for books or lectures, rather than attempting to understand them. She regards many ufologists as egocentrics who take abductees and their experiences and shape them to fit *their* needs and *their* belief systems.

A central concern of Patsy's is that investigators seem to treat their cases as proprietary, and hardly ever hand a case over to a mental health professional, or even suggest that the person concerned seeks counseling. This is a key point, and it seems to have escaped the attention of all but a handful of investigators that some of the abductees may be suffering from problems such as Post-Traumatic Stress Disorder—a recognized psychological condition resulting from exposure to harrowing events outside an individual's usual range of experiences. This is an unfortunate mistake, because those who accept the reality of the abduction events should be the *first* to recognize that such events will have inevitable psychological repercussions.

Patsy believes that the central difficulty is that a ufologist would not necessarily know when or if it was appropriate to refer a case to a professional. She also points out that while there are dangers in not referring a seriously disturbed person to an appropriate professional, there are equal dangers in persuading someone that they need help when they do not. There is clearly a fine line here, and Patsy is not convinced that ufologists even acknowledge such a line, let alone know where to draw it. She makes her views clear:

"Ufologists are interested in aliens, psychologists are interested in people."

Although I think that Patsy goes a little too far in some of her criticisms, there is much in what she says that the UFO lobby would do well to take on board. As things stand, Patsy would never consider regression hypnosis with a ufologist, or permit more detailed examination of her case. There are important lessons here for all those who study abductions, because until researchers gain the confidence of people like Patsy, many interesting experiences will never be reported to anyone associated with ufology.

While Patsy has an intellectual grasp of the alien abduction hypothesis, she regards it as no more than one of a number of possible interpretations of the experiences reported by so-called abductees. While she is prepared to look at such theories, she regards them as being one piece in a very complicated puzzle. On a purely intellectual level she is prepared to acknowledge the possibility of alien abductions, but she does not accept that proof of this would present us with some magical solution to the nature of the universe.

Patsy's interests are inclining increasingly toward Eastern mysticism and the way in which these ancient philosophies seem to tie in with some of the recent discoveries being made in the field of quantum physics. Far from being at odds with ufology and abduction research, such ideas may have a real part to play in helping us understand the abduction mystery. They are not too far removed from those of researchers like John Mack who believe that the Western scientific paradigm is simply too rigid ever to explain fully the abduction phenomenon. Whatever is going on, Mack believes, undoubtedly *manifests* itself in the physical universe, without necessarily *originating* in it. It may originate in an entirely different reality, from a place where Western science is unable to follow. Perhaps this is a place that Patsy and others have somehow glimpsed.

CHAPTER 10

Tourists

This is one of the most fascinating cases that I have ever investigated, and one that challenges many of our ideas about the universe in which we live. Before I began my official research into the abduction mystery, I would have thought that events such as the ones in this chapter were impossible. I would have written them off as either deliberate hoaxes or the product of a mind unable to differentiate between fact and fantasy. But during my investigations I came across too many cases like this to sustain the view that the abduction phenomenon was all in the mind. For this to be true, some sort of mass delusion on an unprecedented scale would have to be gripping the entire world. This seemed unlikely. What seemed more likely was that cases like this kept coming in because the witnesses were reporting a genuine, physical phenomenon.

Mary (pseudonym) was born in 1962 in the Republic of Ireland. She comes from a large family, and was brought up as a Catholic in a household where both her parents worked. In my experience, people of Celtic origin are generally more open to the concept of the paranormal than the average person, and—perhaps *because* of this openness—are more likely to have experienced paranormal phenomena. It certainly seems as if Mary has a predisposition to the world of the paranormal. Her father has seen ghosts, and both her mother and her maternal grandmother were aware of the *other* world, inhabited by spirits, fairies and similar entities.

There are a number of factors that combine to make this case especially interesting. The first is that Mary does not consider herself to be an abductee at all, even though she is adamant that she has encountered *other* intelligences. This is important, because skeptics often like to imply that being an abductee is rather like belonging to an exclusive club, where membership is actively aspired to. This is certainly not borne out by my own research. In my experience the label "abductee" is viewed as undesirable, and is something that most of the people whose cases I have investigated do not want to consider.

It is also significant that despite an interest in UFOs that is perhaps inevitable, Mary has no real involvement in what might be considered the ufological "scene." If she wanted, she could become the center of attention at a conference, and could doubtless get her picture into the newspapers. But she shies away from any personal publicity, feels uneasy among ufologists, and has asked that I use a pseudonym to protect her identity. She has only consented to her story being used at all because of a desire to help other people who have had similar experiences, by showing them that they are not alone. She also shares the anger that many people feel toward officialdom, which does nothing to research the phenomenon, or help those affected by it. In short, she is not the attention-seeking egotist that the skeptics would have you believe are the kinds of people who report such close encounters.

Mary's experiences, like those of so many who claim contact of some sort with aliens, began in her childhood. I have yet to investigate a single case where the experience reported is a one-time event. I am convinced that those who have such encounters have a relationship of some sort with these *other* intelligences. They are not casually picked out and then discarded, but selected, nurtured, and matured.

A final point about Mary is that she has never undergone regression hypnosis. All the experiences detailed in this chapter are consciously recalled.

Like many of the so-called abductees, Mary has witnessed a number of other paranormal events that—at least on the surface—seem to have little to do with UFOs or aliens. The first such incident occurred when she was about four years old, and involved her little brother, Liam (pseudonym). She recalls watching him stumble at the top of the stairs, and fall. She was very frightened, and feared he would be seriously hurt. But she recalls that her father was at the foot of the stairs, and somehow "slowed him down with his mind." Liam fell in slow motion, did a complete somersault, and landed safely in her father's arms.

Shortly after this incident, another strange event occurred, which again seemed to focus around Liam. They had been playing together, along with an older child. Mary is unable to recall much about this other child, and cannot even remember if it was a boy or a girl. The only detail she does recall is that the child was wearing light clothes. Liam went off with this older child, holding his hand. Her next memory was of the two of them rising up into the air, and flying over a fairly busy road, still holding hands. Mary thought that this must be something that everyone could do, and stepped out into the road. A truck came past, and at first it seemed as if she had been hit. Fortunately, all that had happened was that she had been knocked back by the wash from the vehicle, as opposed to actually being struck. Nevertheless, it was obvious that Mary had been hurt. The shocked truck driver took her back home, and her mother bandaged her leg, which had been cut. Clearly recollections can become blurred over the years: Liam remembers nothing of his flying experience with the mysterious older child, but Mary's mother vividly remembers the day when the truck driver brought Mary back home, and explained how she had "suddenly appeared at the side of the road." This incident, and the earlier incident where Liam was seen to float down the stairs may be connected. It might suggest that Liam was the focus of events and not Mary. This is possible, but Mary's own encounters with strange entities were about to begin.

Like many Irish people, Mary was brought up on a rich diet of folklore. This folklore formed a part of her formal education, and was studied as an integral part of Irish history. A common feature was the existence of the fairy folk. Mary was brought up to believe that fairies were real, and her family always ensured that offerings were made to the little people. When a pot of tea was made, a little would be poured down the sink before any of the family received their share. With food, a little was always left on the plate. It would have been unlucky not to follow these customs, and Mary follows them to this day, even though she has long since moved away from her Celtic roots.

When Mary was about seven or eight, she encountered two strange entities, that to her were the fairies she had been told about from early childhood. It was summer, and Mary recollects that she was ill at the time. (As a child she suffered frequent sore throats that would often result in her being confined to bed.) She was in bed, and had just asked her mother to fetch her a glass of water. Mary heard her mother go downstairs, and turn on the tap. Suddenly, she saw two small heads peeping around the bedroom door. She knew they belonged to the little people because their heads were *below* the level of the door handle. As Mary's mother came back up the stairs, the creature on the right hand side of the door dashed over to join its companion on the left, affording Mary a brief glimpse of it, before both of them disappeared. The movement had been dainty, and Mary described it as "a skip." The creature was short, with frail-looking limbs. It had big eyes, and a pointed chin. It seemed to be expressionless, and Mary described it as seeming neither angry nor playful. As Mary puts it: "They were just observing me."

Mary was less sure about their clothing. She cannot recall seeing any fingers or toes, and feels that the creatures may have been wearing one-piece garments that were stretched tight over their entire bodies. Interestingly, she describes them as being not entirely solid. They seemed almost transparent, as if the light was shining through them. Mary likened their limbs to the stalk of a plant, or a stick of celery, where you can see through the material if there is a light source behind them.

Mary told her mother that she had just seen two fairies outside her bedroom. Her mother's reaction might seem strange to those of us who have had a more conventional upbringing. Instead of telling her to stop being silly, or to stop making up stories, her mother asked her why she didn't invite them in to play. She also asked if Mary had spoken to them. Mary said that she hadn't, but added that she would the next time. It was a promise she was to keep, although not in quite the way she might have expected.

If Mary had simply imagined the whole incident, or somehow misinterpreted some prosaic occurrence, it seems likely that the creatures she saw would match her expectations of what the fairy folk were like. This was not the case. Although the physical description was the same, Mary expected fairies to be playful and mischievous. The creatures she saw were neither.

There is an interesting footnote to this story. In 1987 Mary saw a copy of Whitley Strieber's book, *Communion,* and stopped in her tracks. The artist's impression of

Whitley's alien, which was on the front cover of the book, was identical to the fairy that she had seen all those years ago. It was this incident that first caused her to suspect that some of the incidents from her childhood might not be due to fairies after all.

By the spring of 1992 Mary had moved to London, and on the day in question was working near South Kensington in London. As she emerged through the ticket barrier at South Kensington underground station, she noticed someone standing by the public telephones. The figure was male, and only about five feet tall. He was very petite, had a large head, and was wearing big glasses, which Mary said made him look like an owl. He looked very frightened, and seemed very uncomfortable in his clothes (a light gray suit), which were too big for him. Mary felt that everything about him was somehow "not right."

Eye contact was made, and as this happened, Mary had the distinct impression that all was not as it seemed. She had a very strong feeling about the man, and a thought about him suddenly occurred to her: "You're not from here."

The response, when it came, was both shocking and unexpected. Again a thought suddenly appeared in her awareness, which she was sure had come from him. It was something Mary can only describe as telepathy. He asked for directions, although the communication was more akin to a statement than a question:

"The Science Museum."

Mary tried to respond telepathically, but found herself incapable. She wanted to use words, but found herself instead giving a visual answer, as if seeing a map in front of her. Her directions, although technically correct, did not represent the shortest route. It was, however, the route that Mary herself used. When she had finished, the man turned away, and ran up the stairs. Intrigued, Mary followed, and as he left the station, Mary—who was only about ten yards behind—resolved to keep following him. She emerged onto the street only seconds after the man, but there was no sign of him. He had vanished into thin air.

Mary is convinced that the "man" was actually a "visitor." She does not favor words like "alien" or "extraterrestrial," but is happy to say that the entity was not human, as we know it. In fact, Mary realized this straightaway, even prior to the telepathic exchange. She has told me that she was amused at the creature's nervousness. In fact, her impression was that he had lost his "group"—or somehow come adrift from others—and that the Science Museum was a meeting point.

It strikes me that there is a degree of humor in this story. The idea of aliens dressing up in human clothes is not new (contactees such as Orfeo Angelucci and Howard Menger made claims about such behavior, and the previous chapter described a similar incident involving a group of "aliens" seen in a London art shop), but there is something singularly apt about the idea of an alien coming to Earth and asking directions to the Science Museum. What better place to get a good idea about the level of our technology?

Mary's strange encounter at South Kensington underground station mirrors a bizarre incident that occurred in January 1987, and involved Bruce Lee, an editor at the American publishing company William Morrow & Co.

Lee and his wife were in a New York bookstore when they noticed a short couple flicking through copies of Whitley Strieber's *Communion*, which had recently been published by Morrow. It was difficult to make out much about them, because they were wrapped up in bulky cold-weather gear. But they seemed to be flicking through the book as if they were reading at an unnaturally high speed, and all the while they were making remarks about errors in the text. Lee and his wife went over the couple, and Lee introduced himself. As the woman turned to face him, Lee saw that behind her huge sunglasses were massive, black eyes. Shaken, and overtaken by a sudden desire to leave, Lee retreated. Was this simply a case of a rather overactive imagination, or are there really aliens among us, checking out our technology at ground level, and reading books about themselves? If so, what are people like Mary and Bruce Lee actually seeing? Aliens wearing disguises, aliens having shape-shifted to a human form, or screen memories that somehow alter the perception of the witness, causing them to see a human? Whatever is going on, the fact that these entities are noticed at all suggests that any masking technique being employed is not entirely effective.

On New Year's Day 1996 Mary was to encounter another strange character who she strongly suspected was not human. There was a knock on the door of her house at about noon, and as Mary came downstairs to see who it was, she saw the visitor's face through the glass panel on the front door. Opening the door, she saw the figure was standing to one side of the porch. The figure was male, and was wearing a beige-colored coat or jacket. He had messy, wispy hair, wore large, tinted glasses, and was brandishing an ID card of some sort, which was thrust quite aggressively toward Mary's face. He announced that he was conducting market research for a well-known chain of supermarkets, and asked whether she shopped at the company's new store that had just opened in the vicinity. As it happened, she did, but replied that she didn't, feeling that this would get rid of him. The figure made as if to tick Mary off a list, turned, and walked away. Mary was uneasy throughout this event, and felt decidedly uncomfortable about the visitor. She sensed that the visitor, too, had been uneasy, and felt intuitively that he was not who he said he was, suspecting that the story about market research was a cover for something else. Some might argue that even if this were so, the man was not likely to have been a thief, casing various properties for future burglaries. But why would a thief who planned ahead so carefully make the fundamental mistake of calling at a house on New Year's Day, which would inevitably arouse suspicion?

There were other inexplicable things about this episode. The ID card itself was an oddity. While it had the word "research" at the top, it seemed to have columns and columns of text, together with a photograph on the right-hand side. Mary cannot recall what the photograph was, but is sure that it was not a face. Mary's son, Connor (pseudonym), has said that the picture seemed to be of a city. The hand that held the card was not visible, and appeared to be wrapped in folds of cloth.

At my suggestion, enquiries were initiated with the supermarket chain that the caller had claimed to be working for. The result of these checks was intriguing. The company had not been undertaking market research in Mary's neighborhood, but such a project was being planned, and would start in the next few weeks.

There was a rather sad end to this story. Mary's cat, who had been in the hallway when the visitor called—and was visibly spooked by him—became increasingly skittish in the days after the visit, and disappeared shortly afterward.

On Friday March 1, 1996 Mary heard what sounded like two or three sets of very soft footsteps running down the stairs, followed by the sound of a car pulling away outside. This was not the first time that Mary had heard light footsteps in the house, and on each occasion Connor was safely tucked up in bed. And the sound of a car pulling away put me in mind of a screen memory. Had there been visitors in the house that night who had in fact made off in a rather more exotic vehicle?

Another episode that occurred in 1996 was one of the most spectacular missing time incidents that I have come across. Mary was traveling from London to Newcastle, a journey that she had done a number of times, and that averaged out at eight hours of driving, allowing for a couple of short rest breaks. Mary, together with her friend Brenda (pseudonym), were on the M1 motorway, and were approaching the city of Leeds, for which they had just seen a sign. Mary was at the wheel, and noticed what she thought was a small aircraft ahead of them, quite high up, and flying across the road. It turned toward the car, and Mary had a brief thought that it might be in trouble, and might try to land on the motorway. Brenda, meanwhile, bent down to reach for some drinks that were in a bag on the floor by her feet. She retrieved the drinks and looked up, when what she saw caused her to scream in shock and horror: "Where the fuck are we?"

They were no longer on the motorway, but were on a traffic circle in central Leeds. Mary and Brenda had both experienced exactly the same transition from motorway to traffic circle, without being aware of any occurrence that could possibly explain it. It took them half an hour to get back onto the motorway. Fortunately, on the approach to Leeds, just before Mary had seen the aircraft, they had made a note of the time. It had been 2 p.m. An instant later, as far as their perception was concerned, it was 5:30 p.m., and they were on the traffic circle. They had lost three and a half hours of time, and had no idea what might have happened in the interim. The fact that there were two people involved, both of whom share the same memory, is tremendously important. It seems to rule out any theory about it being a dream or hallucination, perhaps occurring during a rest break. The scenario certainly fits that of many road abductions, although the fact that it happened on a busy motorway in the afternoon may require a reassessment of our ideas about exactly how such events take place without being witnessed by countless other motorists. It might be that some sort of mass hypnosis is involved, with witnesses seeing either nothing at all, or having a screen memory of a car at the roadside, accompanied by a breakdown truck, complete with flashing lights.

Spectacular though this missing time experience is, it pales in comparison to the story told by Mary's paternal grandmother about an incident that occurred in Ireland, in a small village in County Louth, in about 1910. All the children walked to the village school from their homes in the surrounding countryside. Mary's grandmother recalls that on one particular morning she saw two of her classmates walking down the hill on the opposite side of a vale that she passed on her way to school. She waved to the children, and they waved back. The two children failed to turn up at school that day, and a frantic search was launched, without success. They were missing for three days, but on the following morning they walked into the school as if nothing had happened. They had no memory of any strange experience, and no recollection of having lost any time.

The details of this incident are vague, and very difficult to check. But the story was told long before Mary had even come across the concept of missing time or alien abductions. It is, I expect, one of countless such occurrences that are perhaps now recalled only in distant memories. The account will doubtless become just another tale amid the mass of Celtic folklore.

Mary has seen a UFO only once, and that was before Connor was born, when she was crossing the Atlantic in a yacht with her ex-husband, Patrick (pseudonym), and Patrick's brother, Michael (pseudonym). On the evening in question it had been about 7 p.m. when Patrick headed off down below to contact nearby boats on their SSB radio. Such casual chat is fairly standard maritime practice. Mary and Michael were on deck, washing up, when all of a sudden the sky lit up as if it were daylight. They looked to the stern, and saw a ball of light shining straight down at the cockpit, illuminating the area where they stood. They shouted frantically for Patrick to come up, but by the time he did so, the light had gone. It transpired that Michael had seen a light moving backward and forward across the horizon earlier in the day, but hadn't thought to mention it. They tried to contact other ships in the area, but without immediate success. When they did establish contact, two hours after the sighting, it became clear that many other ships in the vicinity had seen the light. The crew of one yacht fifty miles from where Mary and Michael had been illuminated from above had also seen a white light shining on the surrounding sea. Ships at distances of up to four hundred miles away had altered their courses and sailed toward this light, thinking that it might have been a distress flare. Over the next few hours there was constant talk about the various sightings over the radio. It was clear that there had been many witnesses. To the best of my knowledge, none of these UFO sightings (which occurred in international waters) were ever the subject of an official report, although I understand that a well-known author was caught up in this incident, and can verify events.

Mary has had a number of dreams that may be of significance with regard to her experiences. Freud felt that dreams often represented clues to what was going on in the subconscious mind, and it may be that some repressed memories of traumatic events surface (albeit in garbled form) only in dreams. Other abduction researchers might regard Mary's dreams in a similar light to screen memories. In one, Mary dreamed that she woke up (a so-called false awakening), looked out of her bedroom window, and saw dozens of cats sitting on the roof, all staring at her.

Another time she dreamed that she was flying, and saw a spaceship.

Perhaps the most curious dream involved her assisting in some sort of medical examination. She had been caring for various sick people, and had showed them great compassion. A number of people in the room (all of whom were very short) then took her to a taller female, because they were impressed with her, and wanted the doctor to know how well she had done. Unfortunately Mary started to boast, and seeing this the doctor wanted nothing to do with her. While describing this dream to me, Mary made what some might say was a Freudian slip, when she referred to a tall "being," before hurriedly correcting herself and describing the doctor as a "woman." At one point in the dream she remembers going to a small room of some sort, and comforting a distraught woman who was apparently hiding from the doctors.

It may be that there is nothing to such dreams. They are certainly not that unusual. But taken in conjunction with all the other events that have occurred, it is just possible that they represent another piece in the puzzle, and that they are memories (perhaps distorted ones) of actual experiences. But might this last dream of Mary's not tie in with the reports made by some abductees of seeing human "collaborators" on UFOs, assisting the aliens?

The investigation into Mary's case is continuing, but it is a case that seems to bridge a number of separate but possibly related phenomena. Mary has encountered beings who she feels sure are extraterrestrials, but she does not consider herself to be an abductee.

CHAPTER 11

Recharging the Psychic Batteries

Like many of the abduction cases I have investigated, this is one where the subject displays a range of what are generally referred to as psychic powers, ranging from clairvoyance to precognition. Like many (but by no means all) abductees, she is mindful of the adverse reaction that society can have toward those who report such experiences. It is no exaggeration to say that a few hundred years ago such claims might have been treated as evidence of witchcraft, resulting in persecution or even death. While modern attitudes are less extreme, they can still be hurtful, and many abductees have had their sanity questioned, have been called liars, and have been ostracized by family and friends. Consequently, in a number of cases I have been asked to protect the identity of those involved. This is one such case, and throughout this chapter I shall refer to the witness as Jayne. I have also changed a few personal details, to avoid giving any clue to her identity. None of these changes affect the overall nature of this account, and no details of the actual experiences have been altered in any way.

Like many of those who have encountered other intelligences, Jayne's experiences have been occurring for many years. However, like many abductees, Jayne feels that there was one experience that is central to her story, even though there were others that preceded it. The reason this event is so important is that other members of Jayne's family witnessed some of the strange happenings, providing corroboration for her account. Jayne told me that it was this experience that gave her proof of what she already suspected.

Although Jayne cannot recall the exact date that the event occurred, she knows it was a Sunday night in November of 1992. She had gone to bed at about 10 p.m., but was restless, and seemed unable to get to sleep. Sometime later she heard a strange, electronic buzzing sound, which made the hairs on her arms stand on end. Simultaneously, the family's two dogs started to whine. A little later, at around 1 a.m., Jayne became aware of a disturbance in the house, and heard her eldest son, Mark (pseudonym), arguing with Sue (pseudonym), his girlfriend, who had been staying in the house with them. The argument became so heated that Mark almost forcibly ejected Sue from the house. This was out of character, and it later transpired that he had felt a sense of impending danger; a feeling that something was going to happen that he did not want Sue to be caught up in.

Jayne's husband, Tom (pseudonym), had slept through the whole incident. He is normally a very light sleeper, who suffers from mild insomnia, but on this occasion Jayne was unable to rouse him, despite shaking him quite violently. After this disturbance, the lights were turned off, the doors closed, and the household settled down once more. Jayne glanced at the clock, and noticed that the time was 1:30 a.m.

At 1:45 a.m. Jayne became aware of a sound that she could only liken to the blast of air heard when an automatic door opens. She now believes that this was some sort of "dimensional happening," as if a portal to *somewhere else* had been opened in her room. She was also aware of a deep rumbling sound. At this point she was lifted up, and felt herself being carried. Jayne's recollection of these events is clear, although to this day she is unsure whether she saw this visually, or psychically. She is able to recollect the images, but feels that for at least part of the time she had her eyes closed. However, I will continue to use visual terms to describe Jayne's memories, for the sake of convenience. By now Jayne was aware that she was no longer in her bedroom, but was being taken down a corridor. She could see little gray figures all around her, and described them as having wrinkled skin, "like a mushroom." There was a strong smell of ammonia, and of something else that she could only liken to Brylcreem.

By this stage she was not sure whether she was being carried, or if she was walking. She caught sight of a woman with blonde, shoulder-length hair, who was smiling. Jayne was feeling understandably frightened, but was calming herself down by telling herself it was only a dream. At the back of her mind, however, she was perfectly well aware that the experience was *not* a dream.

At this point she entered a room that was bathed in an exceptionally bright light. The walls were curved, but the room was probably more oval than round. Jayne describes it as being a "busy environment," with three pyramid designs on the walls, which she felt might be some kind of insignia. A medical examination of some sort was then carried out, although Jayne does not recall the precise details. She was aware of pains in her stomach, and felt that something was done to her head, and particularly to her ears. At the end of this procedure she was taken to what appeared to be a refectory, where there seemed to be a lot of people eating, talking and relaxing. She was offered something to drink, but again cannot recall the specifics. Her awareness—which up to now had been rather unfocused—began to sharpen, and she suddenly began to appreciate the enormity, and the extreme strangeness, of what was occurring. A sudden thought crystalized in her mind: "My God, they're aliens."

As soon as the thought was out, Jayne felt that the entities realized she was not under the mysterious *control* that had made her so docile and compliant, and they did something to her, placing her into an altered state of consciousness. She saw an image of her mother, who had been dead for many years, and Jayne feels this imagery was deliberately placed in her mind, to convince her that she was dreaming. She then felt a curious pulling, as if she was being sucked into a vortex, followed by a thump, as she hit the bed. The time was now 4:35 a.m., and Jayne spent the next twenty minutes trying, unsuccessfully, to rouse her husband.

In the morning Jayne woke up at 7:45 a.m., and immediately made herself a cup of tea. She lit up a cigarette, and then noticed that she was shaking uncontrollably, and sweating. She started thinking about the events of the previous night, and kept trying to convince herself that it had all been a dream. But for the rest of the day she continued to suffer various physical effects, such as shaking, headaches and extreme thirst. Jayne is usually a very calm person, and not at all nervous, but every time she thought about her experience, a feeling of dread came over her. The family's dogs (who it will be recalled had been disturbed by something the previous night) and cat were also acting strangely, and refused to come into the house at all. All the while, Jayne continued to try to convince herself that the experience had not been real, but at 5 p.m., when Mark came home, his first words finally dashed her hopes of passing the incident off as a vivid nightmare:

"What the bloody hell was going on in your bedroom last night?"

A chill came over Jayne as Mark described what he had experienced the previous night. It transpired that after he had made sure that Sue had left the house, he had returned to bed, but had been unable to go to sleep. He had then seen a bizarre green, fluorescent light shining all around his door, as if there was an extremely bright light source just outside his room. He tried to move, but was overcome by a feeling of total paralysis. He too had heard the eerie whooshing sound that his mother had heard, together with what sounded like very loud voices, although quite unlike anything he had ever heard before. He was aware of nothing else until the morning.

Jayne was profoundly affected by this corroborative account, and was very emotional over the next few days. She was afraid to go to bed, and could only do so if her pets were in the room with her (perhaps because she sensed that they were sensitive to the intrusions, and would serve to provide early warning of any repetition). This behavior was out of character, as Jayne is normally confident, and self-assured. The experience also seemed to unlock memories of earlier incidents, and Jayne began to reassess many other strange things that had happened to her before.

Jayne comes from a large family, and from the age of three she shared a room with one of her sisters. She clearly recalls many occasions when a strange figure stood at the foot of their beds. They thought that the mysterious stranger was Jesus, because they had no other frame of reference to describe a person who could magically appear in this way. A little later, when they were told about someone else who could enter rooms by magic—Santa Claus—they believed they had seen him too. They were luckier than most children: not only did they see the lights from Santa's sleigh in the sky, but he came to visit several times a year, often coming into the room, just as "Jesus" had done. This strikes me as an interesting example of children interpreting the phenomenon according to their limited understanding of the world. Very young children would not be likely to have any knowledge of alien abduction, but in a Western, Christian society will almost certainly be familiar with the figures of Jesus and Santa Claus. So any mysterious visitors to their rooms may well be perceived as one or other of these benevolent characters.

Jayne also recollects playing with a black-clad man when she was very young, whom she believes to have been a spirit of some sort. There is a possible link between this sort of experience and an alien abduction, not least because of the possible connection with the Men in Black controversy. "Men in Black" were mysterious figures frequently seen after UFO sightings, and often alleged to have visited witnesses, demanding that they stayed silent about their encounters. Some suspected these figures were associated with the government; some believed they were Walter Mitty types who enjoyed dressing up and intimidating witnesses; while some theorized that they themselves were extraterrestrials.

Jayne's recollection of any later childhood experiences is hazy, and indeed the next two experiences that she remembers in any detail did not take place until November 1991, just one year before the event that was to unlock her memories of these earlier experiences.

In the first incident, Jayne had woken with a start and had been aware of what she thought was an exceptionally bright full moon outside her window. Standing by the chest of drawers was a tall figure that she assumed to be her husband. She wondered why he was standing there, but assumed it had something to do with his insomnia. A few seconds later, the terrifying truth dawned on her, when she moved her arm and felt that Tom was sleeping soundly beside her. It was only the benevolence emanating from the figure that prevented her from screaming out aloud. As Jayne's night vision began to take effect, she was able to get a better view. The figure was male, tall, with long blond/brown hair and possibly a beard. He was kindly looking, and was not dissimilar to the popular image of Jesus. He was wearing a shiny garment of some sort that almost looked as if it were made of polyethylene. It appeared to be a uniform, and there were epaulettes on the shoulders. The figure was leaning against the chest of drawers, with his arms folded in a rather nonchalant pose, and the most amazing thing was that Jayne could see right through him. Jayne has seen a number of ghosts before, but feels that they are mere visual images, and not sentient. Jayne sensed that this entity was very different, and was most definitely self-aware. Jayne hid under the quilt, and eventually the figure disappeared. The following morning, Jayne's daughter, then aged eight, had a question for her:

"Why was Daddy watching me in bed last night?"

Jayne's daughter had clearly seen the same figure that Jayne had done, and had similarly assumed it to be Tom. Tom, of course, had not moved from his own bed. Although Jayne had a positive feeling from her encounter, she was upset at the thought that her daughter was somehow getting drawn into events.

The next strange occurrence took place on a Thursday night, when Jayne had decided to go to bed early. When she awoke, she was not in her bed, but on a very hard surface. There was a warm, comfortable feeling, and she was aware of a lilac/white light above her, in a very narrow

line, as if from a strip light. She was also aware of two figures on either side of her, although again she describes this
as more of an impression than a visual sighting. There is,
as with so many of these accounts, a sort of otherworldly
feeling to events. She then heard an exchange of conversation that seemed to confirm the reality of the situation:

"She's aware of where she is."

"She can't be."

"We've got to take her back."

She was then aware of a spiraling sensation, which
culminated in her return to bed. The next day Jayne assessed the experience. She frequently has lucid dreams
and false awakenings, but was positive that this had been
no dream. It was real.

Aside from experiences that might be defined as alien
abductions, Jayne has also had her share of UFO sightings. The most noteworthy event occurred in July 1992
when Jayne, Tom and a group of six other friends decided
to go on a skywatch, having had their interest stimulated
through various other paranormal incidents. It was a
Friday evening, and the group set up camp on a particular
hillside in the West Country (precise location withheld in
order to maintain confidentiality). As it fell dark they noticed a number of satellites in the night sky, and counted
about eight in the course of twenty minutes. It was Tom
who spotted the pink, glowing object hanging menacingly
in the southwest sky. At first he thought it was a star or a
planet, but then he saw it pulsate, and pointed it out to the
others. They watched it on and off, for the next half an
hour, during which time it was stationary. Then, suddenly,
the object shot off at an incredibly high speed, so fast that

heads and eyes had difficulty in keeping track of it. They had seen military fast jets before, and were used to their high-speed flights. They estimated that this object was traveling at as much as ten times this speed, having gone from horizon to horizon in an instant.

Checks were made with a local military establishment, but nothing should have been in the area at the time in question, and nothing had been detected on their radar. This event was to have a profound effect on Jayne, Tom and the others. They felt that they had been changed by their experience, and became increasingly deep and philosophical about seemingly mundane things.

The effect of the group's UFO sighting mirrors the changes that often take place in the lives of abductees. This has been particularly noticeable with Jayne. She has always believed in the paranormal, which inevitably leads to the usual chicken-and-egg argument about abduction experiences. Does a belief make someone more receptive to an idea, or does the belief stem from the experience— whether consciously recalled or not? Jayne has seen ghosts, had an Out of Body Experience, and developed precognition: the ability to see the future. This ability has grown over the years, and Jayne is now one of Britain's most talented clairvoyants. Her skills are such that in January of 1993 she was approached—very discreetly— by an important Middle Eastern political figure, on the basis of personal recommendations.

Jayne believes there is a definite link between her psychic ability and her abduction experiences. This is a link that I had discovered myself in the course of my own research and investigation, but Jayne epitomizes and articulates it better than most. She believes that her psychic abilities peak immediately after an abduction, and fall back to their normal levels shortly thereafter. She even suspects that this might be one of the main reasons behind her experiences, serving perhaps to recharge her psychic powers in some way. Tom agrees. He feels that Jayne was "upgraded" during her encounters. This hypothesis seems to have been confirmed during one of the two hypnotic regression sessions that Jayne has undergone. (Again, it is important to stress that most of the information in this case was consciously recalled: hypnosis was only used in order to confirm the basic facts, and provide some additional details.) Jayne was in fact quite reticent under hypnosis, but replied to a question about why she had been abducted that it was because of her psychic abilities. She also predicted that open contact would be made in the next few years, with landings in several different locations worldwide. (Such claims are fairly common among abductees, and may stem more from hope than from any firm information; open contact with intelligent extra-terrestrials may result in the abductees' stories finally being believed.)

Jayne's encounters seem to have settled down to a regular pattern of approximately two each year; but that may be a conservative estimate, since she feels that she shuts out much of what is happening to her. She becomes upset and emotional, but does not appear to have suffered any physical harm, so the situation is one that she tolerates. Jayne views the experiences with mixed feelings, seeing both negative and positive sides to her encounters. She believes that the individual's reaction to the phenomenon is the crucial factor. If you face it without fear, you are more likely to benefit from it. She describes how such an attitude can lead to an "uplifting of the soul." She also feels that there is something recognizable about the phenomenon. The uninvited guests who intrude upon her life are familiar strangers. As she has said to me:

"We *know* about this. Is it part of us?"

In the final analysis Jayne's case is a fascinating one, and one that challenged many of my own ideas about the alien abduction phenomenon. It is all too easy to view the whole subject as black and white: either extraterrestrials are abducting humans, or they aren't. But a case such as Jayne's does not fit neatly into any ready-made category. The dividing line between ghosts and aliens, physical abductions and Out of Body Experiences is a narrow one. The phenomenon that we label *alien abduction* may well encompass a number of different elements. The truth lies not in the black or the white, I suspect, but somewhere in the shades of gray.

CHAPTER 12

Terror on the Toll Road

Some of those who have had experiences of the kind mentioned in this book are happy to label themselves as abductees, or at least content to be labeled in this way by ufologists. Some take the opposing view, especially those who believe that use of the word *abductee* creates an instant bias against the extraterrestrials, whom many believe to be here for benign reasons. But perhaps most of those who have been touched in some way by this phenomenon are agnostic in their views. They know that *something* extraordinary has happened to them, but they have no idea what it is. They listen patiently to the skeptics who say either that they've made it up or that they're suffering from some sort of psychological delusion. Then they listen to the ufologists who say they've been swept up in some interstellar breeding program orchestrated by gray aliens from Zeta Reticuli. Somehow, neither theory cuts the mustard, leaving the witness in a somewhat confused state of limbo. Peter (pseudonym) is one such person.

Although there have been a number of anomalous incidents in Peter's life, the only one he wishes to talk about for the moment is the one detailed in this chapter. Peter is British, but at the time of his experience he was going out with an American woman called Jenny (pseudonym), whom he had met while she was studying and living in London. Their encounter with the unknown took place late in the night on January 3, 1991, during a three-week holiday that the couple had spent in Florida over the Christmas and New Year period.

Toward the end of the holiday, Sharon (pseudonym), an old college friend of Jenny's, had come to stay, and the three of them spent three days in the vicinity of Disneyworld near Orlando, visiting, in turn, the Magic Kingdom, Epcot Center and MGM Studios. On the final day they had gone to Universal Studios, and after a full day of excitement they headed back home. They checked out of the Holiday Inn in Orlando in the early evening, stopping at a roadside McDonald's before leaving the relatively built-up area that surrounds the various theme parks.

In order to travel to Boca Raton, which lies on the east coast of Florida, midway between Miami and West Palm Beach, there were two main routes that they could take. The first option was the coastal route, Interstate 95. This would have been a pleasant drive, passing through some of the most affluent areas of America, looking at the brightly lit houses of some of the rich and famous, and enjoying the sight of the ocean alongside them. The second alternative was to take the Florida Turnpike—a toll road that passes through the desolate state interior, skirting around the Everglades and Lake Okeechobee. Although the Florida Turnpike would have been marginally quicker, speed was unimportant, as there was nothing planned for the following day. The toll road was considerably less interesting a route, so it is surprising that this was the one they took, especially as this was Peter's first trip to the United States. There was another reason why the choice of route was surprising: Jenny was nervous, and was prone to panic about crime, and indeed she had recently told Peter and Sharon about how several women drivers had been forced off the toll road and then raped. Despite the apparent advantages of taking the scenic coastal road, the three set out on the more direct, cross-country route.

The toll roads that are so prevalent in the USA work on a very simple principle. You join the road at one of a number of fixed points, and pick up a ticket, either from a machine, or from an attendant. When you leave, it can only be by a number of fixed exits, complete with booths and barriers. You pay on a sliding scale, according to how far you have traveled. There is no way that you can get off the road without passing through an exit booth, and paying. Or is there?

The journey had passed without incident. Jenny was driving the rented Honda, while Peter was at her side, in the passenger seat. Sharon was in the back, and had fallen asleep. Certainly, she has no recollection of the bizarre events that were shortly to follow. Because none of them had any special plans for the following day, and because, as a consequence, they were in no hurry, none of them had made a particular note of the time. But now it seemed to Peter that there was something rather strange going on. He had a map, and had noticed that the distance between the last exit point and the next one was fairly short. The next exit was the one they wanted, and so Peter had been looking out for it on the road ahead. They had placed the ticket they had picked up when joining the road on the dashboard, together with the few dollars that they would shortly have to hand over. But there was no sign of the exit. Just the flat, featureless scrub of Florida's interior.

Peter wondered whether the map might be wrong, or simply out of date, and mentioned this to Jenny. She, after all, was a native of Florida, and knew this route quite well. To Peter's surprise, Jenny had been similarly confused. It was almost as if they were lost, although this is clearly impossible on a toll road with a set number of fixed and clearly marked exits.

Peter and Jenny became more and more puzzled, and then alarmed, as they wondered—out loud—where the exit was. But almost as soon as they had articulated their concern . . . they weren't on the toll road anymore, but were in the outskirts of Boca Raton, looking at the comforting lights from motels, cafés and houses. Peter and Jenny just looked at each other in horror and disbelief. Simultaneously, they hummed the theme tune from *The Twilight Zone*, and this was strangely reassuring. First, it confirmed the experience, convincing both of them that they hadn't suddenly gone mad. Second, it was a useful method of lightening the very tense atmosphere that had developed. As if to reinforce this point, nervous laughter followed their musical comment on the situation. To lighten the atmosphere further, Jenny decided it would be a good idea to switch on the car radio. The radio was pre-tuned to a rock station—one of many that broadcasts to the state of Florida. A track was playing, and after a chord or two of guitar came a lyric from a song that the couple have not to this day identified—perhaps because they'd prefer it that way: *"The girls in The Twilight Zone . . ."*

Peter is unsure what happened next, but is confident that there was a lot of yelling involved!

So what *had* happened, and does this experience tie in with the missing time phenomenon that ufologists see as the primary clue to an abduction having taken place—even if no UFO was sighted? Neither Peter nor Jenny had any knowledge of or interest in UFOs, other than a general awareness that the UFO mystery existed, and that it exercised the minds of various weird and wonderful characters on the fringes of the more conservative and ordered world they knew. Neither of them had seen a UFO prior to the experience, nor had they noticed anything else out of the ordinary. Neither of them had any awareness of the concept of alien abduction, and if you had asked them about missing time they would have simply given you a blank stare. They arrived home at around 1 a.m., but because they had paid no attention to the time, they could not say for sure what time they had left Orlando. They may have lost an hour or two of time, or they may have lost none at all. They simply don't know.

The couple talked about the experience at great length, and agreed that it was the strangest event that had ever happened to either of them. But in spite of this, they did not see fit to mention it to Sharon, or to ask her whether she had been aware of anything unusual. Somehow they both knew that she had slept through the entire event, and had played no part in the experience. This was confirmed some years afterward, when a few vague questions were put to her.

Both Peter and Jenny had made instant and rather unlikely attempts to rationalize the situation. Individually, either of them might have convinced themselves that they had made a mistake. It was dark, they were tired, and the human mind is a complicated device not properly understood by scientists. People hallucinate, and people can be mistaken. But both Peter and Jenny had been through the experience, and a shared delusion of the identical incident seemed too far-fetched—and probably no less paranormal than the event they were desperately trying to rationalize. It was no good trying to explain away one paranormal happening by reference to another. Furthermore, there was the physical evidence that they had not been through the toll booth. For there, on the dashboard, was the ticket and the money that they had prepositioned, ready to hand in.

A bizarre factor in this event was the way in which it seemed to slip from the memory, rather in the way that a dream can be clearly recalled upon waking, but is soon forgotten. Some time after the experience, Jenny returned to America to go to university—a move that was eventually to lead to the couple splitting up, after having made attempts to sustain their relationship despite the geographical separation. On one occasion, no more than a few months after their experience, Peter spoke to Jenny on the telephone and asked her about the night in question. He wanted to check his recollection of events, but didn't want to bias Jenny by suggesting a particular answer, so he simply asked her what had happened. Her response surprised and chilled him:

"Something weird to do with a song on the radio . . . ?"

Peter wondered what was going on. How could she possibly have forgotten something that they had both agreed was the strangest experience of their lives? He reminded her of the core of the experience, something that he refers to not as "missing time" but as "missing road." As if a veil had fallen away, a wave of realization passed over Jenny.

A few months later, Peter was to suffer a similar shock, as he in turn received a reminder that was to bring the experience back to him, after it had slipped from *his* conscious memory. In Peter's case, the reminder came from reading a book he had found in a secondhand shop. The book was called *Perspectives*, and had been written by the respected British ufologist John Spencer; it was about abductions.

There was a story in the book about a road abduction, and somehow this had a profoundly chilling effect on Peter, even though the account started with a UFO sighting—something that had never happened to Peter or

Jenny. For no apparent reason the passage that scared Peter most was a reference to the driver suddenly seeing a signpost for home; and it was this that reminded him of the mysterious events on the Florida Turnpike, which he too had somehow forgotten. The book then mentioned the missing time concept, and an alien abduction scenario followed. By now Peter was feeling extremely ill at ease, and was forced to put the book down for a while. This reaction puts me in mind of just how many abductees I know who have started—but not finished—Whitley Strieber's *Communion*. Peter did finish *Perspectives*, and afterward he was forced to reassess the toll road experience. Ridiculous though the whole concept of alien abduction seemed to him, he was forced to admit that *something* was clearly going on. There was, as he reasoned, no smoke without fire. Might their experience have been an abduction?

Peter put his own experience from his mind over the next few years, and while he became increasingly fascinated with the UFO mystery, he never felt any particular urge to explore what might have happened on the Florida Turnpike early in 1991. The matter might have rested there, were it not for a visit Peter made to America in the autumn of 1995. During this trip Peter was to receive an offer of regression hypnosis, aimed at resolving for once and for all what—if anything—had transpired on the toll road. Under regression hypnosis, a quite extraordinary explanation was finally uncovered.

It took a long time. For whatever reason, Peter's unconscious mind did not want to give up the secrets of that January night. But after gentle but expert coaxing, a story—of sorts—gradually began to emerge. It had started with a general feeling of uneasiness, that insinuated itself subtly into his awareness: "Something's not right."

The next moment the car was being pulled off the road, moving in a gentle, sliding motion as it was somehow drawn diagonally off to the left. The first clue to this had been the way in which the sensation of movement had changed. The feel of the tires rolling over the smooth surface of the road had altered, and Peter was no longer sure that the tires were on the road at all. Curiously, he recalled glancing to his left, and seeing that Jenny was sitting in the driving seat, staring straight ahead, with her hands on the steering wheel. There was a spinning sensation, and a feeling of nausea as the car seemed to be drawn off the road altogether, and then sucked upward, turning as it went. When it stopped, they were surrounded by trees, as if they were in a clearing, a hundred feet or so up in the air. The trees were illuminated by light, but Peter was unsure whether this was simply the light from the car headlights, or from some other source.

Peter glanced again at Jenny, who seemed immobile, like a statue. Somehow this didn't seem surprising, and he wasn't as concerned as he logically should have been. He unfastened his seat belt and swung open the car door, casually telling Jenny that he wouldn't be long. He recalls sitting sideways on the seat, with his legs dangling over the side. Then he stepped out. The only way Peter knew how to describe this was by likening it to various cartoons where one of the characters would walk or run off the edge of a cliff, but carry on moving for a while. As Peter observed while under hypnosis: "Maybe people *can* walk on air after all."

Unlike the cartoon characters, there was to be no spectacular fall to earth for Peter. He was either walking on air, or on some sort of invisible surface. It was firm enough to hold him, but with a little bit of give—a sensation that he described as being akin to walking on a mattress. The next instant he stepped up to avoid tripping on what seemed to be a bulkhead. Suddenly, without being aware of exactly how the transition had happened, Peter was walking down a metallic corridor that he described as being about six feet high and three feet wide. His surroundings were reminiscent of a naval ship, something with which Peter was moderately familiar. It was the bulkheads that made this analogy particularly telling.

On the floor, sealed below the surface, were tubular yellow lights. They were coming on in sequence, and this gave the impression of a moving arrow of light that Peter felt he had to follow. When the lights stopped, he found that there was an open door to his left. He entered this room, which he describes as being rectangular, with what were possibly some shelves on the side. The walls, like the walls of the corridor, were metallic. This was not the shiny metal of cheap science fiction films, but the dull, practical color that one associates with naval vessels. It was functional, with nothing being there for show. There was no light source in the room, but he could see well enough by the light from the corridor, which was shining through the partially open door.

The room itself was piled high with cardboard boxes of various sizes, and the boxes seemed to be full of junk. Peter is unsure exactly what was in these boxes, but clearly remembers that there were several items of the sort associated with road maintenance work. He did some cursory poking about, and recalls seeing yellow hazard lamps and various traffic signs, such as diversion and stop signs. His overall impression was that this was a storeroom, and he recalls a feeling of disappointment. Somehow he was expecting "some great secret," but had ended up with a load of junk. The idea of taking something never occurred to him, and thinking about it afterward he felt that it would have been pointless. There were few other details that could be elicited about this room. Some of the boxes looked quite old, and were held together at the sides by masking tape. The only smell was the smell of the cardboard boxes, and Peter himself could recall no particularly unusual bodily sensation.

Aside from the disappointment that Peter experienced, his main feeling was neither fear, confusion nor curiosity, but amusement. During his regression and afterward he laughed out loud at the ridiculous nature of the whole experience. It seemed so pointless for *someone* to go to all that trouble for no obvious gain. It was as if a trick were being played. Certainly there is a humorous side to some of the abduction accounts I have come across, and the trickster mentality of the fairy folk is well documented.

Upon leaving the room, Peter noticed that the direction of the lights in the floor had reversed, and were now pointing to his right, whence he had come. The urge to turn left and explore further simply never occurred to him. Similarly, he was never once tempted to call out to see whether there was anyone else there. The next instant, he was walking on thin air again, and returning to the car, which was immobile amid the trees, as it had been when he had left it. Jenny was still sitting in the front seat, and Peter does not recall seeing Sharon at all, but feels that she was asleep in the back, slumped out of sight. As he approached the car, he felt that the walk was easier this time. It felt as if he were walking on a firmer surface, and he wondered if this reflected the fact that his awareness now accepted the seemingly impossible act of walking on thin air. A glance behind him revealed only the corridor he had just left, disappearing into the trees.

He got back into the car, closed the door, and fastened his seat belt. At this point, the car was returned to the ground, spinning as it went. Peter recalls that this spinning was by far the most unpleasant part of the entire experience, and he felt physically sick during this final stage of his hypnotic regression session. The car was not put back in the same place from which it had been taken, so instead of being on the straight, featureless toll road, they were on a curved section of highway in the outskirts of Boca Raton. But it was only the fact that they were put down on the other side of the toll booth that alerted them to the realization that something bizarre had occurred, and by the time they were back on the road, there was no conscious recollection of any of the preceding events.

When Peter came out of hypnosis he was asked how long he thought the session had lasted. He felt it had been twenty or thirty minutes, and was amazed to be told it was nearer one and a half hours.

Analysis of such cases is notoriously difficult. Peter himself is not prepared to label himself as an abductee, although intellectually he is aware that this is a possibility that will certainly be suspected by many ufologists. He acknowledges that he had been exposed to information about UFOs and abductions prior to his regression, and accepts that this might have had a bearing on his recollections, which could have been confabulated. But if this were the case, why do his recollections of what occurred differ so significantly from the "standard" abduction scenario? Perhaps because his recollections are only partial; or themselves constitute a screen memory of something even more bizarre. Perhaps Peter was still blocking some memories, even under hypnosis. Be that as it may, Peter is not prepared to say that the corridor and the room were on a spaceship, and as has been stated, he saw no evidence of any entities. Despite this, there is no escaping the fact that if the event occurred as Peter remembers, then he was taken—without consent—and subjected to some truly bizarre experience. There must, surely, have been some purpose behind it.

A fascinating aspect of the case is the way in which Peter seemed to be under some sort of control. He was certainly not acting rationally, and nor was he expressing the fear or wonder that one would have expected. Surely nobody would step out of a car suspended high up in the air without any visible means of support. And what did Peter mean by his casual remark to Jenny as he left the car, when he told her that he wouldn't be long? How could he have known?

There are some noticeable similarities between this incident and the experience of Mary and Brenda on their journey from London to Newcastle, as detailed in chapter ten. Both cases involved a vehicle suddenly being much farther along the road than it should have been, without any obvious reason. In neither case was a UFO or any entities sighted. Yet as someone who researches such incidents, I cannot help but notice their similarities, and feel that some pattern is involved. As I have stated elsewhere, such patterns suggest that there is a real and very physical phenomenon at work, because when independent accounts from all around the world tally—as they so often do in the material associated with abduction research—it suggests that the data comes not from the imagination of the witness, but from individual recollections of a phenomenon that can be frighteningly standard at times.

Peter and Jenny lost touch some time after their split, and it has not been possible to speak to Jenny in detail about the event, although she had earlier confirmed the consciously recalled part of the experience they shared, when they mysteriously left the Florida Turnpike without turning off the road or passing through a toll booth. Perhaps the experience still haunts her, and perhaps strange memories occasionally surface through dreams or in flashbacks. If she has been exposed to material relating to abductions (especially the concept of missing time), it is possible that she will have had the same suspicions about the experience as Peter. But given the way in which the memory had slipped away previously, before Peter raised the subject with her, it is far more likely that she no longer has any conscious recollection of events. Perhaps it is easier that way. But it is frustrating for me as a researcher to realize that the abduction cases we study are merely the tip of a huge iceberg, and that the majority of such experiences are either forgotten altogether, or never reported to anyone.

In the final analysis, Peter and Jenny have a conscious recollection of somehow and mysteriously moving from the Florida Turnpike to the outskirts of Boca Raton, without any apparent transition. We can be fairly sure that *something* strange took place that might be given the general label of a paranormal occurrence. But while Peter's subsequent recollections might suggest an abduction, uncertainties about regression hypnosis must cast doubts on this. And if these events did happen, there is no way to be sure whether they occurred in the physical universe, in the internal, psychological universe, or in some *other* reality about which we can only speculate.

CHAPTER 13

Interview with the Aliens

Vaunda Hoscik has no qualms about calling herself an abductee. She claims to have been abducted by the Greys on a regular basis since the age of fourteen, once being visited every night for eighteen months, before contact became more sporadic. Even though the alien abduction phenomenon invariably seems to involve multiple events as opposed to isolated ones, this does appear to set a record.

Vaunda was born in Dulwich, South London, on February 1, 1975. Her paternal grandfather is Polish, and her maternal grandmother was Romanian; like many of Eastern

European descent, Vaunda is more receptive to the idea of the paranormal world than those who have been brought up in the industrialized West. Indeed, Vaunda claims to be psychic, which itself is important, to this case, as we shall see.

Vaunda has one older brother, and one older sister. She had a hard childhood; when she was young her father had a serious accident, which put a great deal of strain on the family. Vaunda's mother worked, and the children had to grow up very quickly.

On July 6, 1989, when she was fourteen, Vaunda went to bed around 10 p.m., but woke up with a start at exactly 2:46 a.m. She didn't know why she had woken up, but she was aware that something decidedly strange was going on. She was sitting up in bed, and had been overtaken by a curious paralysis that affected every part of her body except her eyes. Suddenly she was slammed back down onto the bed with considerable force, as if by some unseen energy. It was then that Vaunda noticed four gray figures in her room. They were about four feet tall, with large, dark, slanted eyes. They had long, skinny arms, and each hand had two fingers and a thumb. The creatures were not wearing any clothes, but no genitalia were visible. Vaunda tried to scream, but was unable to do so. Her next recollection is of being in a totally different environment, although she has no memory whatsoever of how she got there. (This is interesting in itself, given that so many other abductees do have very vivid images of being floated up into a craft.)

Vaunda found herself in what she described as a metallic, brightly lit room, with banks of computer screens on the walls, and a number of tables, on which other humans were lying. She was confused, disoriented and scared, especially when she felt the creatures touch her, all over her face. At this point, she received a telepathic instruction:

"Don't worry; we only want to help you."

In common with many other abductees' reports, Vaunda originally felt that the instruction had been spoken out loud, before realizing that no words had actually been uttered.

Vaunda was shown a series of numbers displayed on a computer screen, which she could see in a mirror positioned directly above her head. She is unable to recall these numbers, but does remember that the sequence 7, 9, 5, 11 appeared prominently (although she admits she is unsure whether the last digit was an eleven or two ones). She also recalls being asked to memorize sequences of symbols. The symbols were familiar geometrical shapes such as circles, squares and triangles, and there were ten sequences, each containing between five and fifteen shapes. Many abductees and ufologists put great store in the symbols recalled by abductees, in the hope that study of what they believe to be the aliens' language might reveal some clues to their origin and motives. Many abductees recall seeing hieroglyphics similar to those used in Ancient Egypt. Researchers such as Budd Hopkins use these symbols as a means of cross-checking abduction accounts. He deliberately holds back details of some of the symbols on the basis that similar descriptions from other abductees would provide corroborative evidence for the physical reality of abductions. Vaunda does not feel that the symbols she saw were alien. She believes that the beings were using shapes that *she* was familiar with, and that the process was a kind of memory test.

She felt that she had been on the craft for a long time, but when she found herself back in her bedroom (again, she had no recollection of how she moved between the two locations) she saw that it was 3:26 a.m. She recalls being aware that she was sitting on the floor, looking toward the ceiling, with an impression that she had been reading.

Vaunda was tempted to write the whole experience off as a dream, even though it had been far more vivid than any dream she had ever had before. But she had no point of reference against which to measure her experience, so she was at a loss about what to do, or who to turn to.

The very next night, the nocturnal visitors returned. Vaunda was subjected to a similar procedure to the one she had endured the night before, and was again told to re-member sequences of numbers. She recalls that she was constantly told that she need not be afraid, and that the entities meant her no harm. Once again Vaunda's first thought was that the beings were communicating through conventional speech, before she realized that in fact the instruction was telepathic. Interestingly, although she heard the words in her mind, she said that they were flashing past at a high speed, making it difficult to understand. Eventu-ally she had to ask the aliens to slow down.

These encounters continued on a regular basis, and Vaunda's feeling that the process with the numbers and shapes might constitute some form of memory test seemed to be correct. Indeed, as well as testing her memory, it seemed to increase its capacity, and Vaunda's schoolwork improved dramatically.

Unlike many abductees, Vaunda found that the aliens were quite forthcoming about their actions and motives (although, as with the experiences of contactees like George Adamski, we have to consider the possibility that they were not necessarily being truthful). They told her that they came from Zeta Reticuli, and described it as being thirty-seven light years from Earth. Zeta Reticuli is, of course, the star system identified from the Star Map recalled by Betty Hill, and is believed by many ufologists to be the homeworld of the Greys. The aliens also told Vaunda that many other people were taken for testing, and that they generally selected those with some psychic ability. Vaunda is psychic, and will, for example, instinctively know when another member of her family is ill. This is an oft-reported phenomenon, especially between mother and daughter. When Vaunda was recently taken to the emergency room of her local hospital for unidentified stomach pains, no cause was ever found, but it later transpired that her mother too had suffered severe stomach cramps on the night in question. Vaunda has also seen the ghosts of several dead relatives, and has experienced astral projection—the ability to project one's awareness out of the physical body.

The aliens told Vaunda that they regarded humans as primitive, but were prepared to share some of their technology with us. In return, they wanted to rediscover emotions, which they had lost long ago as a result of cloning.

The abductions became more and more frequent, until

at one stage Vaunda was aware of experiences occurring every night for eighteen months. Although Vaunda believes that she was being tested, she regards it as being essentially a two-way process. She believes that the aliens were learning from her; mainly about emotions, although on one occasion she says that she took some books up to show her new acquaintances.

Vaunda had not told anyone about her experiences, mainly because she was sure that nobody would believe her, and that people would think she was mad. This was undoubtedly putting her under a great deal of pressure, and indeed the keeping of *any* secret is extremely stressful. It is difficult to assess exactly what effect this may have had on Vaunda—or on all the other abductees who choose to keep their stories to themselves—but it should be borne in mind that there are inevitably psychological implications in such a course of action. Vaunda had tried to get some proof, and on one occasion managed to take a camera onto a craft and take some photographs. The pictures did not come out properly, and Vaunda was seriously out of pocket, as she had spent a considerable sum on the camera, film and processing. She subsequently found the exact amount of money involved under her pillow!

It is at this point in Vaunda's story that her experiences first began to become traumatic, but the initial trauma came from an unexpected source. Vaunda's improved performance at school had by now attracted the attention of her classmates. She was popular with the older children, helping the fifth-year pupils with their tests while she herself was in the *first* year of her secondary education. Unfortunately, this only aroused the envy of her immediate peer group, who regarded her as a "swot," and resented the fact that she could apparently do so well academically without having to study hard. (It is significant that she obtained nine GCSE passes without ever having applied herself to her studies, and without having done any revision.) Vaunda began to be bullied, and by her own admission became "the class clown" as a defense mechanism. Unfortunately, she also made a conscious decision to slow herself down academically, in order not to arouse the wrath of her fellow pupils. She began to resent the abductions, and to rue the day that she been "chosen."

Vaunda left school as soon as she could, and spent the next five years drifting in and out of various casual jobs. She took courses in gardening and nursery work, but like many abductees she found it difficult to focus on career goals, and found herself doing less responsible jobs than might have been expected for someone of her intellectual level. (This so-called "status incompatibility" had also been apparent in Ken Phillips' Anamnesis Project; referred to in chapter five.) This was Vaunda's own choice, although I suspect it reflected a subconscious fear of doing well, and stemmed from the bullying that followed her academic performance at school. One side effect of this was that Vaunda discovered that when she was not intellectually stimulated, the visitations would stop.

The abductions did not start again until February 1996, and when they did so, it seemed to have something to do with the fact that she had told her new boyfriend, Chris, about her previous experiences.

Chris Martin was born on August 7, 1960, of Anglo-Indian descent. He has one brother, and three sisters. His parents came over to Britain from India in 1955, and Chris was born and brought up in the London Borough of Lewisham, where he had a difficult childhood, mainly because of the racism that was so widespread during the sixties and seventies.

Like many children of his generation, he grew up watching and enjoying the television series *Star Trek*. Perhaps because of this, when he was about seven or eight years old, he had a peculiar recurring dream; he would dream that the sky was full of spaceships, and would awake with a feeling of euphoria. It is possible that this arose purely from his interest in science fiction, fueled by the escapist fantasies that many unhappy children harbor. But it is also possible that Chris himself is an abductee. Budd Hopkins even has some cases where abductees who are within a relationship recall having *first* met their partners on an alien spacecraft! This poses the bizarre notion of extraterrestrials acting as matchmakers—although this is perfectly plausible if you subscribe to the theory that the motive for abductions is to carry out a breeding program of some sort.

Chris was a withdrawn child, with a creative streak. At sixteen, he left school for art college, and it was there that he first found an outlet for his creativity. He took up the guitar, and discovered that he had a surprising aptitude. He became highly accomplished, and enjoyed success as a member of the heavy metal band Thunderstick.

Vaunda and Chris met on November 30, 1995, in a pub where they were introduced by a mutual friend. Vaunda recognized Chris immediately, and claims that this was because the aliens had shown her a picture of their introduction six years before it happened. According to Vaunda, she was highly skeptical about her own experiences when they first occurred, and one day had demanded some proof. She was shown the image of herself meeting Chris, and told that she was seeing an event that had not yet taken place. When it did, they said, she would have her proof.

In a curious twist to their story, Chris had a strong interest in UFOs, while Vaunda did not. In fact, many abductees have no interest in UFOs whatsoever, probably because (consciously or unconsciously) they are so convinced about their own encounters with entities that they feel study of the visitors' vehicles to be irrelevant. They feel, perhaps, that they know more than enough about UFOs already! In February 1996, Chris had a copy of one of the many UFO magazines that have become so popular in recent years, and casually asked Vaunda if she had ever had any weird experiences. Vaunda's reply startled him:

"Well, actually, I've been abducted."

Chris was the first person she had ever felt able to tell about her experiences, even though they had only been together for three months. It was Chris who suggested that Vaunda contact the extraterrestrials again, and although she was initially reluctant, she eventually decided to go ahead, feeling that some good might come out of the experience. She knew exactly how to get back in touch. The couple drove to Vaunda's old family home in Kensington, and parked outside. Vaunda was aware that intellectual stimulation was the key to contact, since it had dried up at her behest when she made a conscious decision to avoid any challenging mental processes. Chris had brought along some IQ tests that he had been given for a job that he was applying for in the computer industry, and Vaunda began to answer the questions with a speed and accuracy that surprised him. Contact was to be re-established, although—as we shall see—matters would soon get out of hand.

Chris plays an important part in Vaunda's story, not least because of his acceptance of Vaunda's experiences and the support that he has given her. Many abductees find personal relationships difficult, perhaps because the person being abducted subconsciously blames their partner for not being able to help them. Despite some initial stresses, the couple's relationship seemed a strong one, perhaps because Vaunda does not have as violent a fear reaction as many abductees. Chris has taken an active and intellectual interest in Vaunda's encounters, helping her record and evaluate events. Perhaps because of this new factor, the experiences themselves began to change, evolving into a phenomenon that is in many ways entirely different from the standard abduction. Essentially, Vaunda's physical abductions seemed to have been replaced by remote telepathic contact with the aliens. This is known as channeling, and involves Vaunda being able to act as a conduit for extraterrestrial information.

The basic idea behind channeling involves an individual entering a trance state, and allowing other intelligences to relay information through them. This either involves the individual writing down the information (so-called automatic writing), or speaking onto a tape. In both cases, it is the other entity who is in control. The phenomenon of channeling is highly controversial, not least because of the difficulty of obtaining any hard proof. Even if the channeler is being entirely truthful, the process is impossible to validate, and could be likened to receiving an anonymous telephone call. One has no real way of knowing who is behind the communication, or judging whether or not *they* are being truthful. In America, an entire New Age industry has grown up around the concept of channeling, and it is not surprising that some charlatans have gotten in on the act.

As with so much of this subject, one has to set aside one's natural desire for scientific evidence when looking at a case like this. You simply cannot prove (or indeed disprove) these claims by reference to any conventional investigative technique. Instead, one has to use one's knowledge of human nature, together with a little bit of common sense. Vaunda and Chris have no obvious motive for duplicity. They are not seeking to make any money from their experiences, and indeed they take a considerable risk in allowing their story to be told. As is often the case with such experiences, people will undoubtedly be quick to criticize and ridicule. Like Mary in chapter eleven, Vaunda is speaking out because she feels she has a responsibility to other abductees. For years she thought she was the only person who was having such experiences. It was a lonely time, and there were occasions when she doubted her own sanity. Even though the subject of alien abduction is better known now than it was when Vaunda was a child, it is still not widely accepted by the general public. From what we know of the extent of the phenomenon, it is highly likely that there are many thousands of people all around the world who are having these experiences, and who—as was the case with Vaunda—think that they are alone. The biggest comfort to such people is often the knowledge that they are *not* the only ones to have these experiences, and Vaunda feels that she can help to get this message across.

There is another reason why Vaunda has come forward, and that is because she believes that the telepathic messages she is receiving are themselves important.

Vaunda had channeled one piece of information before, shortly after her experiences began, and had recorded it onto a cassette tape. Unfortunately, she had little interest in what was said to her at the time, and recorded a *New Kids on the Block* album over whatever wisdom had been imparted!

The information that Vaunda channels comes from a number of beings that she calls Grey Twos. Her main contact is a being she names Antholas, although there is another called Crispin, and one called Minnie. The first two names are approximations (Vaunda could neither pronounce nor remember the real names) and the third, as we shall see, is a nickname. I shall now reproduce some of the information that Vaunda has channeled. Some of this comes in the form of straight prose, and some in the form of an interview, during which Vaunda goes into a trance, and answers questions. The first extract is from Antholas:

> *As the years have passed we have been studying your planet thoroughly. We have come to the conclusion that for every one person who believes in us there are another ten that don't. Therefore we have a harder*

task to convince those of you that don't believe. You have so much to offer us, just as we have a lot to offer you and your society. Hopefully, one day in the near future, we will be able to show you just what we can actually give you. It is an effort to keep ourselves hidden, with your technology becoming more and more advanced. Unfortunately, until we feel we can trust you completely, we unable to reveal ourselves in full. This is why we are taking people, in the hope that they will spread the word of our existence.

Antholas then reveals that around *sixteen* different species of aliens are visiting Earth. He says that some of these carry out experiments on humans, although he claims that his particular species plays no part in such activities. He then states that one reason they avoid open contact is that *they* fear that terrestrial governments would experiment upon *them*. An intriguing hint, perhaps, that their technology is not that far ahead of ours, and that they are fearful of our closing the technological gap. We have, after all, come from horse and cart to stealth fighter and space probe in less than two hundred years. Although Antholas denies that his species carry out experiments on humans, it seems to me that Vaunda's testing might be considered an experimentation. About this, Antholas says the following:

Our purpose is to extend your brains as much as possible, so that you are closer to our level of learning.

Perhaps the most fascinating part of this whole story is the channeled "interviews" that Vaunda is able to carry out.

Consider the following exchange, where the questions are being put by a friend of Chris and Vaunda's, Robin Hewer:

> *Robin: It is said that we humans have an astral body; do you as beings have one?*
> *Antholas: No.*
> *Robin: Do you have another body besides your physical body?*
> *Antholas: No.*
> *Robin: So how are you able to possess Vaunda's body?*
> *Antholas: My mind is in her, her mind is in me.*
> *Robin: So you are saying your mind is transferred to her mind?*
> *Antholas: Yes.*
> *Robin: Are you saying part of your mind is still in your physical body at the same time?*
> *Antholas: No; her mind is in my body, my mind is in hers.*
> *Robin: Therefore some essence of you has transferred to Vaunda's body?*
> *Antholas: That is what I said!*

It seems to me that if Vaunda was a fantasy-prone personality, and was making this up, she would have constructed more creative and elaborate replies. Conversely, if such interviews were possible only because Vaunda was in an altered state of consciousness, eager to please, easily led, and in a highly suggestible state, it is unlikely that the answers would have contained flat denials, or been delivered in an impatient tone that borders upon sarcasm.

Minnie is the nickname that Vaunda has given to the

youngest of the aliens she has encountered, and derives from the mischievous character Minnie the Minx from the children's comic, the *Beano*. Minnie is very young, and is apparently a bit of a prankster. On one occasion when the older aliens were talking to Vaunda about Minnie, they told her: "She's yours."

Vaunda feels this meant that she had been trusted with part of Minnie's education, to teach her about human emotions. But it occurs to me that this simple statement may have another, more intriguing, meaning. Vaunda is reluctant to address the issue of motherhood, although there would doubtless be many researchers who would believe that Minnie is Vaunda's hybrid child.

Like many abductees, Vaunda is not the only person in her family to have had such experiences. When in his twenties, another relation was woken up by a very bright light that seemed to be coming from within the room. At the end of his bed he saw a small, gray figure, dressed in a sort of silver suit. This creature had the large head and large eyes that are so typical of the beings seen in such encounters. Suddenly, from outside the room, he heard the shocked voice of his mother.

"I've just seen a spaceship."

"I'm not surprised," he replied. "I've got the driver in here!"

Such an exchange does more than simply provide some humor amid what can otherwise be a terrifying phenomenon. Perhaps it also serves to illustrate the ability of humans to deal with these experiences whatever they may be.

How can you begin to assess a case like this? Vaunda's

story seems to bridge the gap between the physical world and an unknown territory somewhere beyond. On the one hand, many elements of a "conventional" abduction seem to be present. There are, however, other parts of the story that seem to indicate something more akin to some kind of spiritualism, with Vaunda carrying out the role of the medium. After all, who is to say that the aliens are really extraterrestrials, or that the channeled information comes from travelers from the stars? If Vaunda's visitors *are* extraterrestrial, they are certainly fallible, and subject to very human vices and vulnerabilities. Vaunda has seen them smoke cigarettes, play football and become ill. Is this all part of their attempt to understand human emotions, or are they more like us than many suspect?

It seems to me that it is just as likely that some mischievous spirit is responsible for at least some of these experiences. I can testify to the possibility of this myself, and time spent with Vaunda and Chris has thrown up two or three events that I would certainly class as paranormal (I intend to keep these details secret, and see if similar events occur in any other cases that I am investigating.)

As always, I have to take account of the influence regression hypnosis may have had on the case. Most of the details have been consciously recalled, although Vaunda has been regressed twice by Nicola Dexter, a friend who happens to be a hypnotherapist. The only significant new information to come out of these sessions was an account of some childhood experiences (from the age of about two) that involved Vaunda playing with small figures in her room, who would throw her a ball, and then take it away so that she would have to follow. Chris has learned hypnosis, and has regressed Vaunda on occasion. I am not overly keen on this development (for all the previously stated reasons) but if they feel that hypnosis is a useful tool in investigating and dealing with the experiences, then that is entirely their choice.

Investigation into these extraordinary events continues, and just as I was putting the finishing touches to this book, some disturbing developments took place. Vaunda became convinced that she had given birth to a hybrid child who had been taken from her, and she became extremely distressed. She suddenly became convinced that her experiences were negative, and decided she wanted nothing further to do with the intruding intelligences, whatever they were and wherever they came from. In a short space of time she broke off contact, and destroyed most of the records that she had kept. Inevitably, perhaps, these events affected Chris and Vaunda on a personal level, placing intolerable strains upon their relationship. They have now split up.

With such a case the needs of the investigator must take second place to the needs of those actually undergoing the experiences. If things turn nasty—as seems to have happened here—it is sometimes better to drop the investigation altogether. If we forget this, we end up with a situation where the researchers are no better than the abductors, as they probe the minds of the witnesses, looking for information. Frustrating though it is, we must sometimes forgo interesting lines of enquiry if these might cause psychological harm to the abductee. The truth may indeed be out there, but sometimes we must leave it out there.

CHAPTER 14

The Mud Midgets

Maria Ward was born in 1960, and has lived around Dartford in Kent for most of her life. Her father, Fulop Zoltan, was a Tzegane Gypsy, descended from the old Magyar race. Until she was nine, she spoke only Hungarian. Her mother's descent was a mixture of English, Scots and Irish. She has one younger, and two older sisters.

Maria grew up with an acceptance of the paranormal, although it is important to realize that this was based not on belief, but on experience. Much of this experience revolves around what parapsychologists would call poltergeist activity: Maria can hardly recall a time from her childhood when there was not some kind of strange activity in the house. Interestingly, the poltergeist used to reflect the mood of the children. When there was discord, unpleasant things would occur, such as objects being thrown across the room, or smashed. On one occasion a picture fell off the wall, but only after it had moved upward, to free itself from the hook. Sometimes, when there was happiness in the house, more positive things would happen; a pleasant smell would fill the house, or flower petals would appear, as if from nowhere. Maria christened the poltergeist Fred, after the coal delivery man, who was repeatedly locked in the coal bunker by the entity.

Maria's father used to take his daughters into the woods and teach them about the various plants and animals to be found there. Nowadays this might be called nature studies, but Maria's father called it *Natural Law*. She describes it now as an almost shamanistic experience. Doubtless the skeptics would say that this makes her more receptive to the paranormal. But perhaps it is simply that she was able to see the things that were really there, in the way that many children do before their magical universe is brought to an end by the sort of conventional parenting that denies the existence of the paranormal world. There was, it is fair to say, an *acceptance* of this other world.

The times spent in the woods with her father were happy ones, not least because of the characters that she saw there. One day she asked her father about them:

"What do you call little men?"

Her father replied simply: "Midgets."

From that day on, Maria called the little people that

she saw in the woods Mud Midgets, because they were brown. Later, when her vocabulary had expanded, she referred to them as fairies.

One summer, when Maria was seven, she saw a spherical ball of blue light in the woods, about the size of a football. She knew instinctively that it was "something to do with the fairies." The children wanted to play with it, but for once, Maria's father seemed ill at ease. He went away, and when he came back, Maria noticed that he was crying.

As a child, Maria would often wake up in the middle of the night, and would frequently find herself in the garden. This might have been put down to sleepwalking, were it not for the fact that at the time she was too short to unlock the door, for which in any case she had no key. One cold winter night, again when Maria was seven, she was found in the garden in the early hours of the morning. She had been "playing with the fairies." There was a snowman outside, the snow being that dirty white color that snow becomes. But Maria was adamant that there had been *two* snowmen in the garden.

Maria's mother thought such happenings were evil. On several occasions she carried out the old White Magic ritual of placing a bell, book and candle at the foot of the stairs, the book, of course, being the Bible. The ritual is a traditional way of warding off evil spirits. Maria knew, somehow, that this old superstition was completely ineffective, and could not prevent whatever it was that was going on. But she also knew that her experiences had nothing to do with evil spirits.

Maria had assumed that such occurrences were the natural way of things, and was surprised to learn that they were not. She was convent-educated, and in the early days she got into trouble for even *mentioning* fairies, let alone talking about her old friend Fred. To avoid censure, and to fit in with her classmates, Maria stopped talking about such things. Needless to say, this enforced denial of something she knew to be true led her to rejection of her orthodox religious education, although it did nothing to diminish her thirst for knowledge, which is still important to her to this day.

One night, when Maria was nine, she thought that her mother had come into the room. Maria and her younger sister, Lulu, went out into the garden, although it was 3 a.m. There was a large oak tree, and they were accustomed to seeing barn owls perched on the branches. When Maria told her mother (who, as it turned out, hadn't been into Maria's room that night at all) about this experience, she described what she had seen: "The moon landed in the garden, and the owls were singing from the oak tree."

A few days later, Maria heard a strange ringing noise, and was aware that this preceded her "going somewhere." Lulu had broken her ankle, and was upset because she couldn't go with her sister. The nine-year-old Maria could only describe this experience as "going into the moon."

It was at about this time that the nosebleeds started. There was severe hemorrhaging, and Maria was in and out of the hospital over the next few years. Her nose had to be cauterized on five separate occasions, and finally, when Maria was eighteen, the doctors decided they had no option but to operate, to remove some of the veins in her nasal passage. Immediately after this decision was made, the nosebleeds stopped.

Maria's nosebleeds were not the only medical problem she had faced in her teenage years. Her periods had always been heavy, irregular and painful. She had her first and only child when she was nineteen, and the pregnancy was traumatic. She nearly lost the baby on two occasions, and after the birth it was explained to her that she would not be able to have any more children. When she asked why this was, the doctors said that it was because of a previous laparoscopy that had left adhesions in the uterus. A closer examination revealed a strange scar on her navel, consistent with keyhole surgery. The doctors wanted to know why she had had this operation. Maria was shocked and frightened by the question, since apart from the removal of a cyst from her ear as a child, she had never had an operation of any sort.

Between the ages of eighteen and twenty-eight, Maria's experiences seemed to stop completely. Certainly she has no recollection of any strange occurrences. Life returned to normal, and Maria got on with the business of family life. She decided to broaden her horizons, and started a five-year degree course in philosophy and comparative religion. She was relieved that her experiences had stopped, and although, as we will see, she now has a very positive attitude toward them, she acknowledges that they harmed her for a while.

At the age of twenty-eight, Maria was drawn back into the strangeness that had dominated her early years. There were things going on at the edge of her awareness that she could not quite access. Frankly, she admits that she thought she was going crazy. The experiences were similar to her previous ones, but this time Maria was less able to cope with them. Echoing her thoughts about her convent education, she wondered whether her degree course, enlightening though it was, might be narrowing her perspectives. Worried about the intensity of the fear she was experiencing, she gave up her studies.

At about this time, Maria's son Andras was also becoming more involved in the developing events. A "strange lady" would come into the room to talk to him, and more than once Maria saw him floating above the bed, giggling. This was not entirely unexpected, nor was it the first time that such a thing had happened. On several occasions when Andras was a baby he had been found on the floor of his bedroom, throwing a ball, as if there had been someone else in the room. Somehow, he had gotten out of his crib, although this should have been impossible, not least because Andras was born with a deformity in his foot, and had been in a plaster cast at the time.

Now that Maria's own experiences had restarted, there appeared to be a different focus. Perhaps playing with the fairies is a game for children, or maybe children are better able to cope with such events, perceiving them as more benign. Be that as it may, Maria did not like the new order. On one occasion she had just said goodbye to friends, when there was a blinding blue flash. Next thing she knew, it was morning, and she was lying on the floor by the sofa. Her clothes had been removed, and were lying beside her, neatly folded. She had a splitting headache, and her eyes were sore. Instead of her previous acceptance of such things, this time she tried to analyze what had happened, and arrived at a rational explanation. She considered the possibility that she had been sleepwalking, and had a sudden terrible fear that she might have been walking, naked, through her neighborhood.

In November 1990, Maria's husband John had to go away for a few days to Milton Keynes, for a training course. Despite her strong-willed and independent nature, Maria was apprehensive. For the last two weeks there had been a series of electronic beeping sounds in the house, in a regular pattern: two sets of three beeps. These beeps, despite being very intrusive, had no apparent source. Maria was growing increasingly uneasy, and was tempted to ask her husband not to go for the course. In the end, her pride got the better of her, and she said nothing.

On November 21, her husband went away as planned. That evening Andras had gone to bed at about 9:30 p.m. Maria was restless, and spent the evening watching television. She put the kettle on, ready to make a hot drink, but just as the kettle came to the boil she heard the strange beeping sounds. It was 11:30 p.m., and she was overtaken by a sudden compulsion to go to bed, which she did. At exactly 3:17 a.m. she was shaken awake, as if she had been grabbed roughly. There was nobody to be seen. She got up and lit a cigarette, and it was then that she noticed a bright source of light outside the house, where there should have been darkness. She looked out of the window and noticed something strange that "looked pretty." Whatever it was, it started to move. It appeared to be shaped like a gigantic wheel, complete with spokes, and an outer rim that was festooned with multicolored lights: she could see blue, white and red lights, and they appeared to be rotating around the rim, clockwise and counterclockwise. In the center, at the hub of this "wheel," there was a bright white light, far more intense than any of the colored lights on the object's rim. Maria didn't know what it was, but she somehow sensed that it shouldn't be there.

The object moved closer, and she panicked, moving away from the window. Somehow, she doesn't recall exactly why, she was afraid for Andras, who was sleeping in the room with Maria (a special treat whenever John was away, when he was entrusted with "looking after mummy," making him feel very grown up). She tried to wake him, but without success. She then started to slap him around the face, trying to bring him out of the catatonic state into which he seemed to have slipped. She slapped him so hard that hand marks appeared on his face, but he could not be roused.

Suddenly, a blue/white light began to come through the wall, expanding slowly outward so that the light was always spherical. It was rather like a balloon being blown up, and somehow passing through the solid matter in the wall. At this point Maria "heard" a voice that told her not to be frightened, and instructed her to follow the light. She has great difficulty in describing the mechanics of this. It was, she says, neither an audible voice, nor one that might be described as telepathic. It was almost like having a video recording played inside her mind, but she accepts that even this analogy falls short of an adequate explanation. It is, she says, one of those things that you simply cannot explain to somebody who has not experienced it; like explaining color to someone who has been blind all their life.

Maria did not want to follow the light, which had by this time passed completely through the wall. It was spherical, and about the size of a football. It somehow led her down the stairs, drawing her down against her will. Although self-aware, Maria felt that she had no control over the situation; no option but to obey. Once at the bottom of the stairs, by the locked front door, Maria received another command. She was to bend down, and pick the object up. She fought this compulsion, angry at being coerced in this way, and willed herself not to pick it up. She was still thinking of Andras, and used this thought to try and somehow escape the control she was under, by concentrating on something different. At this point an event occurred that seems to me to illustrate the intelligence, and perhaps the humor, of the forces at work. She glanced downward, and saw not the ball of light, but several bottles of milk. She was struck by a sudden thought that this was safe, and nothing to worry about, so she picked them up, only to find that she had picked up the ball of light. A surge of anger ran through her as she realized she had been tricked; but by then, it was too late.

The next part of the experience is one of the most bizarre events that I have come across, and one of the most fascinating. Maria was somehow led, by the ball of light, through the physical structure of the front door itself. She describes the process as being like "moving through molasses," and recalls that she was aware of the internal structure of the wood, complete with splinters, gaps and knots. This was a very slow operation, but once through the door, she was pulled upward quite quickly. As she went up, she was aware of the trees below her, and although she has never particularly liked heights, she knew she wouldn't fall. She felt "lost," not afraid. Her eye was caught by a tennis ball in the gutter. It was one of those curious details that often proves so vital in separating a physical experience from a purely psychological one, because some weeks later the gutter became completely blocked with leaves. As the guttering was being cleared, Maria remarked that there was a tennis ball up there. There was.

She was feeling slightly dizzy, and there was pressure at her sides and above her neck. She felt that she was not being pushed up, but sucked up. There was a vaguely unpleasant sensation, akin to being in an elevator—and indeed elevators have always been something that Maria has found unpleasant, and avoided, for as long as she can remember.

All the while, there was a strange ringing sound coming from somewhere, together with a sort of low-frequency vibration that was decidedly uncomfortable; the sort of effect that you get from standing near the speakers at a rock concert, where you not only hear the sound, but *feel* it passing through your body.

Maria was suddenly aware that she had stopped moving, and was standing on something solid. Her hands were at her sides, and she was looking ahead. She was still worrying about Andras. The atmosphere was dim and hazy, but she could see shapes moving about in front of her. There were shuffling sounds, which Maria described as like that of children moving around on gym mats, in bare feet. It began to get lighter, although not from any discernible source. All of a sudden the scene became clear, and Maria saw that there were three small, brown creatures standing directly in front of her. She recognized them, and the idea that she had seen them before was somehow the most frightening aspect of the whole experience. Maria is five foot four inches tall, and by comparing herself with these creatures, when they stood next to her, she estimates their height as being between three and a half and four feet. They were colored brown, like mud, and were wearing no clothes. They all looked the same, and there was no flexibility to their movements, so that they walked in an extremely rigid fashion. The creatures had no genitalia, no belly button, and no facial or body hair. Their skin was smooth, except for a little creasing at the waist, shoulders, and the side of the head. They had no discernible ears or nose, but large eyes. Maria is keen to point out that these eyes, although large, were not quite the huge, wrap-around eyes so beloved of ufologists. Somehow, and Maria has no idea how, she knew that the creatures were male.

One of the beings stood in front of Maria, while the other two positioned themselves at her sides, and lifted her arms. A doorway of some sort had opened in front of her, and the beings led her forward. She had to stoop to get through. They veered to her left, and began to walk around what appeared to be a semi-circular corridor. They arrived at a little room, and again, the doorway had seemed to appear in front of them, in a place where previously there had been no opening. Two of the beings left, and for the first time, Maria found that she was able to move her head a fraction. She took advantage of this to have a look around. The room was small, and completely empty, although running all the way around it was a sort of ledge, about a foot wide and some eighteen inches off the floor.

Maria had the impression that this was something designed to be sat upon. The ceiling was high, and Maria recalls thinking this very odd, as the creatures were so small.
Another opening appeared in front of her, revealing a
second, larger room, about thirty feet in diameter. Like the
previous room, this one had a domed ceiling and was virtually featureless, with no joins visible, even between the
walls and the floor. There was another ledge around the
side of the room. Maria now works for the Royal Mail,
and using terms that are familiar to her, she describes this
feature as a "parcel shelf."

In the center of this room was a raised, rectangular
platform, about six feet by four. Maria sat on the side of
this, legs dangling over the edge. She recalls being surprised that although the table was silver-colored, and
apparently metallic, it was not cold. It was neutral in
temperature, and had the feel of plastic. The room was
quite light, but again, there was no visible source of illu-
mination. She noticed that there were five of the small
brown creatures in the room with her, and heard a
whooshing noise behind her, and to her right. At this point,
she thought she was going to die, and was sad that nobody
would ever find her, or know what had happened.

Maria saw movement out of the corner of her right
eye, and another creature then came into view and stood
directly in front of her. Although she felt that the creatures
were essentially the same species, this being was physically markedly different from the others. It was about the
same height as Maria, and was a sort of dirty white color,
the way that porcelain can look when it has aged. Its face
was slightly fuller, and there was a small bump that might

have been a nose, and a slit where the mouth should have been. The eyes protruded more, and were larger than those of the other beings. They were deep black, with a trace of movement beneath the surface. Once Maria had looked into its eyes, she was unable to look away. The being somehow communicated a thought to her:

"Has it been so long."

This was not a question, but a statement, delivered in a very matter-of-fact manner. It lifted its right arm, and touched Maria's shoulder. Maria then lay down on the raised platform, and almost immediately felt what she could only describe as thoughts being somehow pulled out of her. As she told me: "I didn't feel like my soul was my own. It was belittling. It was like someone was raping my mind."

Although the process was painless, she stressed that she had not consented, and it was this that made her angry, as opposed to what was being done to her. Maria felt that she knew what was happening, however. She felt that her emotions were being tested.

The little brown creatures were scurrying around rapidly, as if at some sort of command, and Maria felt that she was being held down on the table, although she was not aware of any restraints. One of the creatures then produced an instrument that looked like a short rod of about a foot in length, and stood at the left hand side of the platform. The device was a lime green color, and when viewed from directly ahead, appeared to be triangular. At its tip was a bright green light. The being poked Maria with this rod, all over her body. Although this was not painful, bruises subsequently appeared on her body at each point that the device had touched her. When it touched her arm, she was able to see inside her skin, viewing the flow of blood, and even her bone structure.

The being then cut the fingernail on the little finger of Maria's left hand, and she remembers being particularly annoyed at this, as she always grew her nails quite long. Something was stuck into her finger, and although the skin was not broken, a tiny circle of red dots was visible for eight weeks after the event. Maria was telling herself that the experience was a dream, but all the time she knew that it wasn't.

A rectangular block began to descend from the ceiling, and Maria was overcome by the terrifying thought that she was about to be crushed. But the block, which seemed to be an exact match of the one she was lying on, stopped just above her face. A bright, blue/white light shone down at her out of this structure, and moved up and down her body. Although the light was bright, she could feel no heat. Her head was turned to the left, and she felt a tug at the back of her neck. She saw that one of the brown beings was holding a small, flexible filament, about ten inches long, with a bright light at the end. The being moved around behind her, and inserted the object into her neck, just below her right ear. At this point Maria felt the most terrible pain she had ever experienced in her life. Again, she thought she was going to die, and felt a sort of acceptance, as she went beyond pain. She was able to move a bit, and shot out her left hand, grabbing the tall being's wrist. She shouted at the creature: "You can't!"

She received a sort of instant feedback through touch, and formed the distinct impression that the beings had no understanding of the concept of pain. The tall being touched her forehead, and stared into her eyes. At this, the pain evaporated, and Maria felt very calm. It was as if she was being anesthetized. Maria was crying, and could feel the salty taste of tears that were dripping off her nose and into her mouth.

Maria glanced downward, and to her horror saw something moving *inside* her stomach. There was a sharp pull at her navel, and then the sound of things being put in some sort of tray, somewhere at the foot of the platform, out of sight. The sound was akin to metal cutlery being thrown into a plastic container. At another stage in the procedure, something was placed up her nose. Maria recalls a noise rather like the sound when you break the seal of a new jar of coffee.

One curious fact about all of the beings was their scent; a slightly pungent smell, similar to rotting leaves. This smell was noticeable throughout Maria's experience.

Finally, all the beings left, apart from one of the brown ones. Maria was now able to move, and realizing that she was naked, put on her baggy T-shirt, which was lying beside her, neatly folded. She had no recollection of how or when it had been removed.

Once she had put the garment on, she realized that the last of the brown beings had gone, leaving her alone in the room. Maria began to count, to try and get some idea of the passage of time. She had reached the number 256 when she heard a sound, turned to her right, and saw that another one of the taller beings had somehow come into the room. Maria instinctively knew this was a female, before she noticed that the being was wearing what she could only describe as a crude, paper dress. It was a simple, triangular affair, and Maria felt that it was unnatural, and being worn not out of any need, but just for show.

The being stared at Maria, who felt the scrutiny to be uncomfortable, and intrusive. She went through a complete range of emotions, good and bad, and had the distinct impression that they were being induced. This procedure then evolved into one where she was aware of images flashing in front of her. Again, some of these images were pleasant, while some were highly distressing. Afterward, she felt emotionally drained.

The most striking image had been an apocalyptic one. She had been viewing the Earth, as if from space. But it was not the beautiful image that we know from NASA photographs, but a dirty, swirling mass of clouds. Hanging over the North Pole was what appeared to be a horrendous bruise, yellow and black in color. Could this have been a warning about the hole in the ozone layer over the Arctic? Suddenly there was a ringing sound, and the Earth fell in on itself. She saw a separate image of a nuclear explosion in the South of France, near the coast.

Perhaps the most intriguing image was that of a desert, orange and brown in color. There was no wind, but the atmosphere seemed to be filled with dust. There was a glow in the sky from what she assumed to be the sun. She looked up, and through the gloomy atmosphere could make out what appeared to be *two* orange suns. She felt this to be a very important place, and it triggered a strange childhood memory. Once, when she was seven, she had painted a picture depicting square huts, with square window-like openings, underneath a sky that contained twin suns. Could this be an image of the aliens' home planet, orbiting a binary star system?

There followed another apocalyptic image, this time of flood waters covering the Iron Age hill fort of Bratton Camp in Wiltshire. There were other images, played as if on some sort of three-dimensional television screen that had somehow incorporated her into the action. Some of the images were of global events such as wars, while some were of more personal images, from her own life.

As suddenly as the process had begun, it ended, and Maria sensed that this had indeed been the climax of the experience. She turned to the being, and an exchange occurred that by now will be familiar to readers:

"I'm never, ever going to forget this."

"You *will* forget it."

A wave of tiredness enveloped her as she was led out of the room and back around the corridor. She was within a light source, and next thing she knew, she was walking back up the stairs in her own house, although she was aware that there was someone or something with her. Aware that she was now on the bed, she glanced at the clock. The time was 4:23 a.m., and Maria suddenly felt extremely sick. She had a headache and a stomachache, and her nose and ears were sore. Her feet were freezing. She pulled the duvet completely over herself, and went to sleep.

When Maria woke up the following morning, she felt terrible. At this time, she had no conscious recollection of the events of the previous night. She went to the bathroom, and when she looked in the mirror, she noticed dried blood around her nose. There had also been a little blood on her pillow. While brushing her teeth, she caught her own reflection in the mirror, and had a flashback of somebody watching her. She also had a memory of a bright light. When she brushed her hair, she noticed something very strange: quite a bit of hair was missing from the back of her head. At this point, she started to cry. There were further revelations in store. Her T-shirt was covered with a strange orange material, and her feet were dirty. She felt as if she was in shock, and had a number of flashbacks. It was then that she noticed the remains of the cigarette that she had lit when she got out of bed. This was another clue that was to lead to her near-total recall of events, by convincing her that the previous night's events had not been a dream.

Over the next month or so, memories came back, in disjointed bits and pieces. While Christmas shopping at a department store in Gravesend, Maria saw one of those lamps with cascading filaments, looking rather like a sea anemone. This brought to mind the image of the instrument that was inserted into the side of her neck. A bonfire, held after Christmas to burn all the boxes and wrapping paper, gave out a smell of musty cardboard, and brought back a memory of the smell of the beings. Maria became physically unresponsive, and couldn't bear to be touched. This was unusual, because she and her husband had always enjoyed a good physical relationship.

Another significant change in Maria was a sudden, profound and totally unexplained claustrophobia. Maria had never had claustrophobia before, and the condition became so debilitating that she decided to consult her doctor, Dr. Pitt. He felt she was suffering from traumatic stress, stemming from some intense and unpleasant experience. When conventional questioning failed to turn up any obvious cause, the doctor suggested that Maria undergo hypnosis, for the purposes of relaxation.

A hypnotherapist—Dr. Susan Winne—was found, and she suggested that she regress Maria to try and find any incident in her past that might have led to the condition. Often, recall of the initial (usually childhood) trauma results in the condition disappearing. Imagine the hypnotherapist's shock when instead of a conventional account, such as the subject having been shut in a small, dark cupboard as a child, Maria recounted the story of how she had only recently passed through the structure of her front door. She recalled that the sensation was like moving through molasses, and felt that she would suffocate.

At this point, Maria spontaneously regressed to a time when she was sixteen years old. She had been in the woods, when somehow—she doesn't recall how—she was put into an oval-shaped capsule of some sort, which appeared to be made of Perspex. She was sealed in, and escorted toward some object in the woods, which she described as "the moon." She could see nobody, and recalls that it felt not as if the capsule was being carried but as if it was being floated along, a few feet above the ground. There was a curious weightless sensation about the experience, rather like the feeling one gets in an elevator. The capsule filled up with a sort of gelatinous pink liquid, and again, Maria felt that she would be suffocated. At this point the hypnotherapist ended the session, and refused to carry on, saying that she was simply not able to evaluate the material that Maria was coming out with.

It is important to stress that this is the only such session that Maria has undergone. Once these memories surfaced, the rest began to emerge unaided, slowly but surely. Maria estimated that she has recalled about seventy-five percent of her 1990 experience, but does not wish to risk the potential distortion that might result from further regression. She does use self-hypnosis, but only for relaxation purposes, and never to try and unlock further memories. She is pragmatic about this, happy to let the memories either emerge by themselves, or not at all.

Although Maria's 1990 experience was the most profound that she can recall, there have been other, more recent incidents. It is possible that these were just as rich in detail as the one I have just described, but, as yet, the full memories have yet to emerge. What Maria does recall of one such incident is fascinating.

Late at night on July 28, 1992 Maria was in Wiltshire, and was driving out of the village of Alton Barnes (famous as the place in which some of the world's most spectacular crop circle formations have appeared), toward the village of Avebury, which lies inside Europe's largest prehistoric stone circle. Suddenly, a dense patch of mist appeared in front of her, and as she drove into it, the engine began to die, and the car slowly came to a halt. A large orange light appeared in front of the car, at which point the mist completely disappeared. Blocking the road, about fifty feet in front of the car, was a large, egg-shaped object, about thirty feet high, and sixteen feet across. Maria stayed in her car, waiting for something to happen. After five minutes, she got out and walked slowly toward it. After she had walked about halfway toward the object, she could walk no farther. Her hair was standing on end, and there was an overpowering smell of ozone in the air. When she stopped, the light came toward her. She managed to shoot off a few photographs, but unfortunately they are far from clear, probably due to the object's rapid spinning and intense brightness.

Maria can recall little else of this experience. Her next memory is of being back in her car, in the village of Avebury. She felt that something profound had happened, but could not recall any details. She did not regard it as negative.

On another occasion Maria's car was followed by a small blue ball of light. This time there were two other people in the car with her, Edward Sherwood and Simon Cooper. All of them witnessed the strange light.

A potentially more disturbing incident occurred on June 5, 1995; but again, Maria only has a vague memory of events, and does not wish to risk the potential contamination that can result from regression hypnosis. All she can recall is looking at what she describes as things that were "like dolls . . . but not dolls." There were a series of what looked like large, glass jars, rather like those used in home brewing. In these were small humanoid creatures that seemed to be alive. They were perhaps six inches in length, and transparent; she could see veins under the skin, which looked wrinkled, and seemed very old. Maria was not sure what she was looking at, but she found it very unpleasant. It went against what she calls her "womanly sense."

To this day Maria has various experiences. She is now interested in UFOs and crop circles, although she is dismissive of some of the wilder theories. She shies away from words like *alien*, *extraterrestrial* or *spacecraft*. She does not claim to be an abductee in the conventional meaning of the word, regarding this as being an assumption that cannot be backed up by any hard proof. Maria views the creatures as alien only in the sense that they are alien to our general experience and understanding. Ironically, given her conscious recall of these events, she probably has less belief in the Extraterrestrial Hypothesis than many ufologists.

What are we to make of a case such as this? The skeptics might simply say that Maria was lying, but she has passed two polygraph tests, administered by a qualified expert, Jeremy Barrett. Furthermore, she has no motive for not telling the truth. She has never sought to exploit her experiences by making any money out of them, and indeed, like Vaunda and Chris in chapter thirteen, takes a considerable personal risk in allowing her real name to be used in this book. But, like them she feels she has to speak out, if not for herself, for the many others out there who have endured a silent hell, undergoing these encounters, while all the time thinking they are going mad. They are not mad. And they are not alone. There are countless people like Maria all over the world, each of them trying to come to terms with their individual experiences.

Maria has certainly given her experiences a lot of thought, and has tried to draw some conclusions of her own. She believes these other intelligences have always been with us, and thinks that we play an active part in this interaction, but without knowing "the rules of the game." She says that they are *not* strangers to us, and that at some level we recognize them. She acknowledges the transformative nature of the experiences, and thinks that the entities might be waiting to guide us through a forthcoming evolutionary leap forward. She believes that if anything is truly frightening about these experiences, it is the way in which they invalidate so much of the conventional "wisdom" that we are taught.

Maria is critical of closed-minded attitudes, whether they are skeptical or credulous. She refutes the assertion that abductees "want to believe," and probably questions her experiences as much as Philip J. Klass would do. On the other hand she is insistent that the experiences—whatever they are—are real. Denying them, she says, won't make them go away.

Maria is now working with other people who have had similar encounters, and is trying to help them in any way she can. Perhaps this is the important point. If we can't stop these experiences, then all the study and evaluation in the world might make no difference. The central issue may not lie with the nature of the phenomenon at all, but in our reaction to it, both collectively and on an individual basis. Maria was once told that life was about two things: love and knowledge. She believes these are the keys to interpreting her experiences. A more positive and constructive attitude is difficult to imagine.

CHAPTER 15

The Usual Suspects

When we consider the bizarre nature of what is commonly termed the alien abduction phenomenon, it is no surprise that an equally bizarre collection of theories has emerged to try and account for the information that has been collected. Taken at face value, the answer is very simple. The conventional explanation favored by many ufologists is that the events described are occurring pretty much as alleged by the abductees themselves. In other words, extraterrestrials are taking people against their will, and carrying out various medical procedures, most probably with the aim of creating alien/human hybrids.

But many people are not convinced that the abduction phenomenon is extraterrestrial in origin at all, and in this chapter I will be addressing one fundamental question: if abductions are nothing to do with an extraterrestrial presence, what *other* theories might explain what is going on?

The following explanations are offered by way of an alternative to the Extraterrestrial Hypothesis. The list is by no means exhaustive, but is aimed at highlighting other factors—some obvious, some not so obvious—that might explain the abduction phenomenon. Clearly none of these theories will explain the entire phenomenon, but just as most UFO sightings have prosaic explanations, many abductions are likely to have conventional causes. I very much doubt that there is one solution to the whole mystery, and I believe there are probably a number of *different* things going on at once, so the following should simply be regarded as a series of phenomena that could be responsible for *some* abduction reports.

Birth Trauma

This theory emerged from the "imaginary abductee" experiments mentioned in chapter seven, in which Alvin Lawson and William McCall attempted to see whether there were differences between the experiences reported by abductees and those constructed by a control group who had not been abducted. There were comparatively few differences, and while many felt that this was because the control group were likely to have encountered at least the basics of an archetypal abduction through media reports, Lawson and McCall arrived at a different explanation. They looked for a common factor that might link the two groups, and were drawn to the fundamental experience of birth itself.

It is certainly the case that some of the images reported in a typical abduction experience mirror images and concepts associated with birth. Many abductees report being in round rooms, in an environment that contains no right angles. Might this not be a distorted memory of the womb? The beings themselves, of course, bear more than a passing resemblance to the appearance of a fetus. The central procedure reported by so many abductees is a medical one, and is recalled as traumatic. Is it not conceivable that recollections of abductions are actually recollections of birth? The white rooms so often reported are reminiscent of a hospital, and abductees have even coined the phrase "doctor being" to describe the entity who is in charge of the procedure. Many psychologists believe that the initial trauma of birth is one of the most profound experiences of our lives, and is largely responsible for shaping the human psyche. It would hardly be surprising if this event—which will clearly be a focal memory in our subconscious mind—resurfaces from time to time, albeit in disguised form.

The problem with such a theory, however, is that birth is common to all of us, whereas the abduction phenomenon seems to be experienced by only a small percentage of the total population. Are the abductees simply those with better recall of childhood events (this could be checked), who are more likely to have some memory of their own birth—albeit a distorted one—or is the entire birth trauma theory incorrect?

Near Death Experiences/Out of Body Experiences

There is always a danger in trying to explain one mystery in terms of another, but it is worth having a brief look at Near Death Experiences (NDEs) and Out of Body Experiences (OBEs), in view of the fact that some of the sensations and images associated with these two phenomena are similar to those reported during abductions. In a typical NDE, which is said to occur sometimes in hospitals, when a person has clinically died for a few seconds, the subject reports a lifting sensation, after which their awareness actually leaves their physical body. There is a feeling that they are traveling down a tunnel, at the end of which is a bright light. There is an impression of great love, and often images will be reported involving the appearance of departed family members, a beautiful environment such as a garden, or even religious figures such as Jesus. Maybe this *is* what happens when we die, but obviously our accounts of NDEs come exclusively from people who were subsequently revived by hospital staff. These reports often describe how the person was told that they had to go back because their time had not yet come, or contain accounts of the subject being pulled back toward their body, as if they were attached by a long piece of elastic that had brought them back with a jolt. Some medics and psychologists believe such experiences arise as an automatic consequence of the brain being starved of oxygen; but this would not explain the cases where those who reported NDEs recall details of conversations that occurred outside the room they were in at the time.

The OBE is a similar concept to the NDE, but is not limited to occasions where the subject is critically ill or even clinically dead. In a typical experience the subject will be lying in bed, when their awareness leaves their body, and floats upward. They will typically report seeing their body lying below them, and this is often accompanied by a feeling of panic, caused by a fear that they will not be able to return to their body. The return, when it comes, is again invariably accompanied by a jolting sensation.

Some people claim to be able to deliberately project their awareness out of their physical form in some way and travel to other locations, or even other planes of existence. Such ideas of "remote viewing" or "astral projection" are beyond the scope of this book, but the commonality of sensation and imagery between abductions, NDEs and OBEs (especially the lifting sensation and the idea of traveling toward somewhere bright) is nevertheless remarkable.

Hoaxes

There is no getting away from the fact that hoaxing does go on. As is the case with UFO reports, some abductions are simply made up. This happens for a variety of reasons, and it is worth examining these to gain a better understanding of some of the difficulties that abduction researchers face.

Some hoaxes will be carried out because the individual craves fame and fortune. It is an undeniable fact that there is money to be made from books, films and interviews (especially in America), and the fame that goes with such high-profile reporting can itself be alluring. There are clearly many people who will do anything to get on television (witness the crowds that inevitably form whenever television cameras are set up in public) and constructing an abduction experience with yourself as the central player is just one of many ways to get yourself onto a talk show.

Some people hoax for the fun of it, and because they enjoy fooling the experts. Witness the exploits of Doug Bower and Dave Chorley, who were responsible for many of the crop circle formations that so intrigued the mass media for much of the early nineties. Although I believe that some crop circle formations are *not* hoaxes but have some other cause that is not yet understood, there is no escaping the fact that Doug and Dave and their imitators were responsible for numerous patterns. They claimed that apart from enjoying the beauty of an English crop field late at night, a key motivation was watching the experts getting ever more excited and devising progressively more exotic theories. Other hoaxers refer to their creations with pride, using phrases such as "land art." Abduction researchers beware!

In addition to those who are motivated by greed or amusement, there are people who hoax as a result of psychological conditions. The concept of the "fantasy-prone personality," for example, is recognized by psychologists. Such individuals have a rich fantasy life, and may be more liable to make up an abduction experience (or misinterpret something more prosaic) because it places them at the center of dramatic events outside the scope of their real lives. Walter Mitty—fictitious creation of the writer James Thurber—is perhaps the most famous illustration of the way in which such characters behave.

Hoaxers are often highly intelligent people who are capable of constructing intricate and believable fantasies. Even the best of researchers can be deceived by such people. One "abductee" whose case was investigated by John Mack has now admitted that she made the whole story up, and if someone of Mack's experience can be taken in, so can any of us. This is something that any researcher has to accept, and indeed it would be naïve of any investigator to suppose that he or she would automatically spot a hoax. We forget how complex some hoaxes are, and just how devious some people can be. For example, there have been numerous cases of people posing as police officers, medics and even train drivers, where lives have been put at risk by the individuals concerned carrying out duties that they were not qualified to undertake; yet the perpetrators were able to maintain their deception for some considerable time before detection.

False Memory Syndrome

Memory is not an accurate record of past events. We all make mistakes, but sometimes even memories about which we are certain turn out to be a misinterpretation of what actually took place. Official police training stresses that eyewitness statements must be treated with caution, because of the way in which the human mind is a fallible and selective recorder of events.

Sometimes when we play events over in our minds or relate an experience to someone else, what started as something about which we were unsure finishes up as a certainty. This is called False Memory Syndrome, and may be a key factor in turning data that emerges from regression hypnosis into a memory that becomes increasingly real as the story is repeated. We have already discussed some of the dangers of regression hypnosis, and it is entirely possible that a mixture of previous knowledge about abductions, coupled with confabulation and False Memory Syndrome, could produce a concrete "memory" of an abduction when no such event had in fact occurred. The American psychologist Robert Baker showed how easy it was to implant false memories in people using hypnotic suggestion. In a famous experiment, his subjects were soon "recalling" occasions when, as children, they had been lost in a shopping mall, even though it had been previously established that this had never happened.

Hypnagogic and Hypnopompic Imagery

Just before we fall asleep, and just after we wake up, our minds are in a curious state of consciousness. In this state we can sometimes see imagery that seems far more real than what we see in dreams. This imagery is known as hypnagogic where it occurs prior to sleep, and hypnopompic where it occurs on awakening. In 1996 the *British Journal of Psychiatry* reported the results of a survey involving over 5,000 people, which found that around thirty-seven percent of the sample reported hypnagogic hallucinations on a regular basis, while around 12.5 percent reported similar experience of hypnopompic hallucinations. The survey reported that the most common image or sensation was that of falling. (This mirrors the standard scenario of many nightmares, as does being trapped by fire, another frequent hallucination.) Of particular relevance to abduction researchers is the fact that the second most commonly reported effect was the feeling that someone or something else is in the room, and the sensation of flying is another typical illusion that may have some bearing on the abduction phenomenon.

Temporal Lobe Lability

The human brain is incredibly complex, and even today we are only beginning to understand how it works. The brain is generally acknowledged to be the seat of our consciousness and our thought processes, and the stimulation of various parts of the brain can produce some curious changes in the way in which we perceive events.

The temporal lobes are areas of the brain particularly associated with memory, emotion and our understanding of speech. These areas of the brain can, to a variable extent in individuals, be extremely labile. That is to say, they are unstable and responsive to stimulation. Psychologists such as Michael Persinger and Susan Blackmore favor the idea that people with labile temporal lobes may—when their temporal lobes are stimulated—experience sensations and/or hallucinations that might be misinterpreted as an alien abduction. These symptoms can include the impression of a strange presence, a sense of ascending into the air, and a feeling of panic—exactly the sort of things reported by the abductees.

Persinger is a psychologist based at Laurentian University in Canada, and has recreated some of these effects in his laboratory. He uses a modified motorcycle helmet to deliver magnetic field pulses to the brain, firing the temporal lobes. While filming a program on alien abduction for a 1994 episode of the BBC2 program *Horizon*, Sue Blackmore tried this out for herself. Blackmore is senior lecturer in psychology at the University of the West of

England, and was keen to see whether some of the factors commonly reported in alien abductions could be recreated. But while she did experience a strange pulling sensation, coupled with spontaneous feelings of anger and fear, there were no UFOs, no aliens and no on-board medical examination.

Electromagnetic Effects

Some researchers suggest that many UFO and abduction experiences have their roots in electromagnetic anomalies that can cause hallucinatory experiences in those who come into close proximity to them. Researchers such as Paul Devereux believe that such electromagnetic anomalies might be associated with tectonic stresses within the Earth's crust. These might manifest themselves as balls of light (so-called "earthlights"), which could be reported as UFOs. And such electromagnetic activity was bound to have an effect on people too, since it would stimulate areas of the brain—in particular the temporal lobes.

Michael Persinger correlated UFO reports with areas of high earthquake activity and found a match. Like Devereux, Persinger believed that the tectonic stresses associated with earthquakes could generate electromagnetic fields, which in turn gave rise to luminous atmospheric phenomena that might explain UFOs, and to hallucinations that might inspire abduction accounts.

The researcher Albert Budden has developed a similar theory, which he calls the electro-staging hypothesis. He believes that electromagnetic effects act as a catalyst for abduction experiences. The electromagnetism may stem from piezo-electrical charges that build up in areas of tectonic stress, and might be released by either earthquakes or by less severe tectonic strain. The resultant discharges may manifest themselves as luminous phenomena such as earthlights or ball lightning. As well as such natural causes, Budden believes that the electromagnetic effects that he sees as the key causative factor in reports of alien abductions may be generated artificially by radio-wave transmitting masts and the sorts of high-tension power cables that are carried by pylons. Over the years, Budden has made vigorous attempts to correlate UFO and abduction experiences to the presence of such structures, and has claimed some success.

The second strand to Budden's theory deals with the way in which these causative electromagnetic stimuli give rise to altered states of consciousness, which he believes occur because the unconscious mind—fulfilling a need to "express" itself—uses such effects to generate the abduction scenario. This scenario emerges from the belief system of the witness, which can differ according to factors such as culture and upbringing. The details will therefore vary, and Budden believes that electromagnetic effects trigger internally produced experiences that might be interpreted as alien abductions by some, but as ghosts, angels or other manifestations by those with different belief systems.

Sleep Paralysis

Even if we do not recall it, all human beings dream. This simple fact can be verified by monitoring neurological activity during sleep. The reasons we dream are not fully understood, but it is believed that some dreams may be the brain's way of assimilating experiences and memories (especially new ones), while some serve as an escape mechanism so that we can give an outlet to fantasies that would be inappropriate or dangerous if they were carried out while we were awake. Other dreams may have something to do with the mental process of problem-solving.

Many dreams involve experiences that simply do not occur in "reality," such as flying like Superman. Another commonly reported dream is one where the subject is being pursued by something, but is only able to run away in slow motion.

Sleep paralysis is that process which inhibits our movements during dreams, stopping our bodies from acting out the motions of the dream. Where this paralysis is ineffective, sleepwalking occurs. But in the borderline state between being asleep and awake (the same state that gives rise to hypnagogic and hypnopompic imagery) the process of sleep paralysis can still be present for a moment, and can lead to situations where people wake up momentarily unable to move.

In a borderline state of consciousness, where a person is just waking up, tries to move and finds they can't, is it any surprise that they sometimes form the impression that there is something holding them down? As one would expect with any genuine phenomenon, sleep paralysis has been known to us for a very long time. But in the past it was often interpreted literally, and many cultures contain stories of supernatural beings that visit people at night. The incubus is the name given to a demon who seduces women in their sleep, while the succubus is the female demon who seduces men.

Interestingly, and in an even more direct parallel with the supposed reproductive aim of alien abductions, visits from incubi and succubi were said to lead to the births of demonic children, or children with supernatural powers.

The hag is another common figure in European folklore. Hags are traditionally ugly, although, like succubi, they are capable of appearing as beautiful for purposes of seduction. Typically, the hag will straddle her victim's chest as he sleeps or lies in bed, and it is easy to see how sleep paralysis could have given rise to such accounts. Canadian folklorist David Hufford has catalogued hundreds of accounts of such a phenomenon, but the "Old Hag" stories that he amassed from Newfoundland in the seventies were almost certainly the modern descendants of legends that go back to at least medieval times, and probably much earlier.

The Collective Unconscious

It is not widely known that the eminent Swiss psychologist Carl Jung made a study of UFOs, and tried to explain them in psychological terms. Even many ufologists are unaware of his 1959 book, *Flying Saucers. A Modern Myth of Things Seen in the Sky.*

Jung believed that humans share a common store of race memories, which are not consciously recalled. Unlike personal memories, which could be suppressed or repressed on an individual basis, these memories are buried deeper still, within all of us. Jung called this database the "collective unconscious," and claimed that its existence could be deduced from a study of instinctive behavior, through which humans were somehow tapping into the sum experiences of their ancestors.

Jung believed that through the collective unconscious we inherit certain basic, primordial images, which are common to us all. Jung called these images "archetypes," and these include the square, the circle and the wheel. Jung argued that some UFO reports might be a modern manifestation of the way in which people can tap into the collective unconscious and retrieve archetypal images such as the circular mandala.

Jung died in 1961, before the concept of alien abduction had surfaced, but it is interesting to note that several of the Jungian archetypes are human or semi-human forms, and encompass a number of mythological figures, such as dwarves. So on the basis that Jungian psychology suggests that some UFO sightings are generated by archetypal images from the collective unconscious, an equivalent argument could be made for the abduction experience, with particular reference to the small beings habitually associated with the procedure.

Jungian psychology is itself highly controversial, and even Jung himself admitted that a psychological explanation alone seemed inadequate to account for the significant number of UFO sightings correlated by radar.

Mass Hysteria

Hysteria is a complex psychiatric disorder with a range of symptoms that include hallucination. But psychologists and psychiatrists are unclear as to whether hysteria is a single condition or various different conditions, and a number have argued that hysteria does not exist at all, except in the minds of various doctors. Such a point is remarkably similar to the criticisms that skeptics make about abduction researchers!

Mass behavior is another psychological term that denotes the collective behavior of large groups of people, where influences spread through wider mediums than personal communication. A good example might be the way in which a fashion could spread as a result of a particular style of clothes being worn by a charismatic character in a popular television series.

It is possible that alien abductions might be generated by a combination of these factors. The mediums of communication would be both science fiction and the abduction literature itself. Science fiction has certainly featured the idea of alien abductions for longer than many suppose. As long ago as 1935 an edition of the US science fiction magazine *Astounding Stories* had a front cover showing an alien looking down at an immobilized human on what looks like a medical table. The number of science fiction films in existence today is phenomenal, and many feature the concept of either alien invasion or abduction. It is possible that these ideas, which are clearly now familiar ones, have been picked up by people who perceive the experiences as real and personal through some "hysterical" condition that psychologists do not yet understand.

Child Sex Abuse/Ritual Satanic Abuse

One of the most distressing crimes that human beings commit is the sexual abuse of children. Clearly such events have a devastating effect on the victims, who will carry the mental scars of this trauma with them for the rest of their lives. But under some circumstances the human mind acts as a sort of censor, wiping out or suppressing memories of particularly traumatic events. In psychological terms this is known as repression. Essentially, it is a self-protection device by which the traumatic memory is expunged from the conscious mind, and buried deep in the subconscious. The problem with this is that these memories never truly disappear (all our experiences and memories are stored in our subconscious), and can resurface at any time. But they may not surface in a logical and coherent manner. They may emerge in jumbled or disguised form, and if the individual concerned is familiar with theories about alien abductions, might they decide that they have been abducted, rather than abused?

Some people have pointed out that the feeling of helplessness experienced by abductees, together with the gynecological focus to the procedures that are commonly reported, are not too dissimilar to the sort of experiences and emotions that someone who has been sexually abused might report. This theory is unsubstantiated by any research into the prevalence of abduction accounts among those who acknowledge that they have been the victims of childhood sexual abuse, and such research seems to me to pose huge ethical problems about just how far researchers should go in trying to evaluate the abduction phenomenon. If such research is undertaken at all, it is something that should be carried out only by appropriate professionals, and certainly not by unqualified ufologists.

One of the few ufologists who possesses sufficient professional standing to look into this theory and make qualified judgments is John Mack. Mack has not come across a single case that has convinced him that childhood sexual abuse—or indeed any other trauma—lies behind abduction reports. On the contrary, as was mentioned in chapter seven, he believes that several cases of apparent childhood sexual abuse were in fact distorted recollections of alien abduction experiences.

Those who believe that childhood sexual abuse might explain some abduction accounts may also regard ritual satanic abuse as another childhood trauma that might surface in a similar fashion. But here it is necessary to strike a particular note of caution. While it is an undeniable fact that some people practice black magic, and some people sexually abuse children as part of organized pedophile rings, many believe that ritual satanic abuse is a modern myth, unsubstantiated by any hard evidence, and inspired by a mixture of sexual abuse, black magic, False Memory Syndrome, and simple exaggerations and lies.

Government Research and Mind Control

As far as I'm concerned, these two separate but interrelated theories go off the scale of paranoid delusions. But as long as such allegations are made, then I feel duty bound to highlight them in this discussion of alternative theories.

The first theory suggests that abductions are an elaborate cover for experiments carried out on various individuals by their own governments, or by some other nameless but firmly human agency. The whole business with aliens is simply an officially sanctioned hoax, designed to sound so ridiculous that it drives away serious investigative reporters.

The second theory derives from the first, and states

that the abduction phenomenon is the result of mind-control experiments, again carried out by governments on their own citizens. While a weapon that could induce such powerful hallucinations would undoubtedly be of great military value, the theory fails for a number of reasons, including the global nature of the phenomenon, and the fact that the subject is now so high-profile that the abductors—whoever they are—are the subject of increasingly serious and sophisticated investigation. This is hardly a situation that governments would allow to continue if they were carrying out such experiments.

Time Travelers

This is cheating slightly, because this theory differs from the Extraterrestrial Hypothesis only by virtue of the fact that the protagonists are not the same. The concept of abductions taking place in a physical sense still holds true. Under this theory, the abductors are not aliens, but human beings from the far future (or even from an unknown technological past). The idea of humans going back in time to change their own destiny is a popular theme in science fiction, and has spawned some interesting paradoxes, the most common being that of the traveler who goes back in time and kills a grandparent, posing questions about whether such an act would cancel out his or her own existence. The time travel hypothesis theorizes that humans in the future have suffered some genetic calamity that they are trying to correct, and that the only way they can do this is to make alterations in the distant past (i.e. our present), without our gaining knowledge of our future, which might lead us to attempt changes ourselves. There is a seductive attractiveness to the theory that the Greys are the creatures into which we ourselves evolve. It would explain why they are essentially humanoid in appearance, and it makes sense that as we evolve, we become more cerebral and less physical as technology lightens our load. Is it not likely that our physical evolution would reflect these developments, with our heads becoming larger while our bodies grow more spindly?

There is, of course, no evidence for such a hypothesis. It is no more than a theory, albeit an intriguing one. It is far from clear that the future has any *existence*, as we would know it; and while some scientists are beginning to look at the concept of time travel as at least a theoretical possibility, to explain abductions in terms of time travelers is another case of trying to unravel one mystery by reference to another.

The Shared Earth Theory

Like the time travel theory, this idea accepts abductions as a physical reality, but lays them at the door of a different agency. The theory has its roots in the folklore outlined in chapter one, but regards such stories as a history of our interaction with entirely *terrestrial* intelligences.

It is possible that humans are not the only intelligent life form on Earth, but that we have always shared it with others. These are the fairy folk of European legend, and the jinn of Arabian mythology. They are variously described as mischievous, and endowed with the power to change their shape at will. It is entirely possible that abductions may occur in the physical sense, but that the abductors are alien only in that they are beyond our understanding. Perhaps we share this Earth with creatures that evolved alongside us, from the same genetic stock. Perhaps we have a symbiotic relationship with these creatures, and if so it raises the bizarre possibility that *everyone* is abducted, and that the process is just as necessary to us as dreaming. The phenomenon may be one that is only recollected by a few, and then misinterpreted through reference to modern cultural factors such as science fiction.

Beyond the Unknown

Science is heading into uncharted waters. Scientists studying the bizarre world of quantum physics are uncovering a region of paradoxes, where concepts of time travel and parallel universes are freely discussed, and where the result of an experiment is dependent upon the experiment being observed. Even the idea of reality itself is no longer sacred, as validated experiments seem to suggest a world in which objects only exist as probabilities until the act of observing brings them into a more solid being. In such a world, anything is possible, and the idea of abductions being explained in terms of *our* "reality" interacting with other "realities" must be considered. If abductions are taking place, then it is possible that those responsible might come from another dimension or a parallel universe.

There are certainly some compelling reasons to suppose that abductions do *not* take place entirely in our reality. The lack of independently witnessed incidents (with the spectacular exception of the Linda Cortile case) is one such clue. We have plenty of photographs and videos of UFOs, but I am aware of no such material showing a human being ascending a beam of light into a craft. Another difficulty with a literal, physical interpretation of the data is the logistics of abduction. If these events are as frequent as, for example, the Roper Poll implies, then thousands of abductions would be occurring every day. While I accept that the UFO phenomenon is widespread, such figures would mean that our atmosphere is *infested* with UFOs, and we might reasonably expect to see even more convincing evidence than we have already.

Summary

I have covered a lot of ground in this chapter, and illustrated a number of theories that might explain some of the many thousands of accounts of abductions that have emerged over the years. Some of these theories are scientific, while some are at least as bizarre as a literal interpretation of the phenomenon. But it is this literal interpretation that I am inevitably drawn to, both by reference to my own research and investigations, and through interpretation of the work of others. While the theories in this chapter may explain *some* abductions, I am convinced that they do not explain the vast majority. The reports seem too detailed, and too similar to be anything other than an accurate description of a fairly standard procedure that takes place in the physical universe. I believe that extraterrestrials are taking humans against their will, into craft, and doing something with them. In the next and final chapter, I will work from this assumption, and address the key questions. What are they *really* doing, and to what end?

CHAPTER 16

Looking for Answers

I believe that careful study of the available data on abductions shows beyond a shadow of a doubt that there is a genuine phenomenon of some sort at work. Although, as we have seen, there are a variety of theories in circulation concerning the precise nature of this phenomenon, I believe there to be an extraterrestrial explanation. My own research and investigation coupled with careful analysis of the work of others leads me to the conclusion that a literal interpretation of the reports from witnesses is the correct one. Put simply, thousands of people have reported being taken into a craft, against their will, by alien beings.

It seems to me that the most likely explanation for this is that the witnesses are, generally speaking, describing events that actually took place. This final chapter works from the assumption that extraterrestrial beings *are* abducting humans, and assesses a number of theories that attempt to explain what is happening.

Study and Experimentation

Some people believe that extraterrestrial visitors to the Earth are so far ahead of us that they regard us in the same way as we would regard the animals with which we share this world. Just as we capture some of these animals, study them in laboratories, tag them with an electronic locating device and then release them into the wild, so there are those who believe that we are on the receiving end of an almost identical process. The parallels are indeed striking, even down to the idea of an implant that allows the subject to be traced, facilitating recapture at a future date.

As always with this question it is dangerous to apply our own human logic to explain behavior that may follow a totally different set of rules. But our logic is the only tool available to us, and although it may not be adequate, we have nothing better. Applying this logic one could speculate that any intelligent extraterrestrial civilization might well be curious about us, and would probably want to study us in some detail. This raises moral concerns about the issue of consent, and abductees will clearly resent any implication that they are being treated in a way reminiscent of the way in which our scientists treat laboratory rats. Time and time again I hear abductees say that if only the aliens would ask their permission, they might actually consent to much of the activity that takes place. The abductees would do this for altruistic reasons, such as a desire to establish a meaningful dialogue that would benefit both parties.

The problem with our protests about consent is that they assume a level playing field that may not exist. The issue of consent does not seem to trouble those humans who carry out experiments on animals, but you don't have to be a genius to work out that rabbits aren't particularly keen on smoking cigarettes or having shampoo put in their eyes. Put bluntly, vivisectionists know that their experiments cause distress, yet they justify them by pointing out benefits that they believe make their work a necessary evil. The same might apply to abductions. The protests of the abductees may be understood but ignored by the aliens, either because they regard protests in the same way that those who experiment on animals regard the squeaking of laboratory rats, or because they are deriving benefits that they believe justify the whole process. As a final thought on this theory I would suggest another parallel with the way that we treat animals. Perhaps the aliens observe our factory farms and vivisection laboratories and assume that to us such things are normal and acceptable. Perhaps we are simply being treated as we ourselves treat others.

The Puzzle Theory

This is a variation on the experimentation theory, but one in which the whole abduction phenomenon is regarded as being a test to measure human intelligence. We are like the rats in a laboratory experiment, scuttling around a maze in search of food. This theory also embraces the UFO phenomenon; together, perhaps with a whole range of paranormal activity. The phenomenon is regarded as a puzzle, but one that gets increasingly difficult. We start from the basic idea of UFO sightings, and evolve through the contactees and the abductees. What makes this theory particularly interesting is speculation about how close we are to solving the puzzle. And if we do solve it, might it not be replaced by something even more bizarre?

Nothing to Do with Us

The human ego is a curious thing. We are so wrapped up in our own importance that we always believe we are at the center of everything. So it is with the abduction phenomenon, which we believe revolves around us. But there is a school of thought that says that while the phenomenon is entirely real, it is in fact nothing to do with us. This theory was developed by the researcher John Spencer is his book *Perspectives*, in which he offers it as an alternative to the Extraterrestrial Hypothesis. Some of John's innovative thinking led him to the idea of our sharing the Earth with a culture that we fail to understand; and he suggests that abductions might be messages from such a culture, which we misinterpret. To me the most interesting part of his theory is the concept of our misunderstanding a process that was not designed with us in mind at all. John illustrates this concept with the following analogy.

If a family of moths—who for the purpose of the analogy are assumed to be capable of some limited reasoning—lived close to a highway, they would see cars and trucks, and sometimes humans. They would see the highway lights, and would be attracted to them. Occasionally one would die on the windscreen of a vehicle. Would the moths not reasonably assume that some complex and mysterious process was going on, in which they were victims of the unknown machinations of some alien intelligence?

Let us take this idea a step further, and draw a closer analogy with the UFO and alien abduction phenomena. At some state in the history of this colony of moths, the first sighting of a car will have taken place, and an excited moth might have related events to the disbelieving colony. Eventually, the sightings of these strange, metallic objects would increase, until the day when a car pulled up on the hard shoulder and a human emerged. Such occasional entity sightings continue, until one day a moth is sucked into a car through an open window. The moth sees strange creatures, who may well be waving their arms about and trying to knock the moth back out through the window. This eventually happens, and the moth, dazed and confused, returns to its colony and tells of how it was forcibly taken on board a strange craft, how it interacted with the beings inside, and how it was finally returned after what was a traumatic experience that took place without the moth having consented to the process.

At first, the moth will probably not be believed, but as a broadly similar experience happens to an increasing number of moths, perhaps some moth researcher will start to ask questions about the phenomenon, and wonder why moths are being subjected to such a bizarre procedure. The answer, of course, is simple. Looking at this analogy from the outside it is perfectly clear to us that the involvement of moths is incidental to the whole scenario. The moths have been caught up in events that are nothing to do with them, and are way outside their comprehension. It is entirely possible that the abduction phenomenon is nothing at all to do with us, and that we have simply been caught up in events that are beyond human understanding.

The Breeding Program

This is the most popular variation on the Extraterrestrial Hypothesis as applied to the abduction phenomenon, and it is easy to see why. Time and again we encounter reports from the abductees themselves that seem to point inexorably to this interpretation. So many accounts—some from conscious recall and some that have emerged under regression hypnosis—tell almost exactly the same story, pointing to the fact that we are being used in a breeding program aimed at producing human/alien hybrids. So much of the very standard experience seems to bear this out.

A medical procedure of some sort runs as a central theme through so many of these accounts. There is a gynecological focus to this procedure, added to which we have reports of eggs or sperm being taken from abductees. Finally there are the accounts of missing fetuses, strange creatures stored in jars on the craft, and presentation of children that seem to involve listless hybrids who—on occasion—the aliens have said belong to the abductee concerned. The case files of the likes of Budd Hopkins, David Jacobs and John Mack are full of accounts of such procedures, and seem to leave little room for doubt that the aliens are using genetic material from us to create a hybrid race. But all too often, that is where the theory stops, as if this were a complete and final explanation. It may be an explanation of *what* is happening, but it ignores the far more important question of *why*.

If extraterrestrials are creating hybrids in a situation where we are not being consulted, then it seems fairly clear that this is not likely to done for our benefit. We are regarded as a source of useful genetic material, but little more. From descriptions of the Greys, it seems that they are cerebral, but lacking in emotion. Are they the dying race beloved of science fiction writers? Have they lost all the raw, primitive emotions that we have in abundance? For all our faults, we have a mixture of curiosity and determination that has taken us out of the Stone Age and into the Space Age in a matter of a few thousand years. Is this what they want from us?

Another theory about hybridization is that it is a prelude to invasion. Some researchers believe that while the Greys themselves may be unable to live on Earth, they are creating creatures that will be able to do so.

Tempting though it is to settle for these theories about the breeding program, there are some legitimate questions to be raised, casting doubt on the validity of the entire hypothesis. The most obvious question relates to the need for a protracted and complex program of abductions. If the extraterrestrials were simply after genetic material, why not take it, en masse, from terrestrial laboratories? One big "hit" could keep them in sperm, ova and other genetic material for years. It seems that the aliens go to a lot of trouble to acquire material that could be obtained far more easily elsewhere. Unless the genetic theory is wrong. It might be that we are misinterpreting the data, or it might be that we are taking the bait and falling for a neat but incorrect solution designed to distract us from the *real* answer to the mystery.

Changes

The modern UFO mystery began with Kenneth Arnold's June 1947 sighting of nine unidentified craft, and was closely followed by the Roswell crash. From then on, the phenomenon appeared to expand exponentially, both evolving and growing in intensity. Sightings were followed by contacts and then by abductions. UFO sightings (and encounters with other intelligences) had been reported throughout history, but why did the pace of things heat up so much at this particular time?

Imagine that you are an extraterrestrial civilization, and that your attention has been drawn to Earth by the radio signals that have emanated from our planet for many years. It is perhaps a reasonable assumption that your first priority would be to conduct a reconnaissance of this newly discovered world, and try to evaluate the sort of society that exists there. But reconnaissance in the forties will have revealed that large parts of the world were caught up in savage and bloody conflict. More detailed observations will have revealed the following specifics:

—The Holocaust.
—Deliberate bombing of civilian targets.
—The dropping of two atomic bombs.

—The development of rocket technology in the V-2
  weapons.

The first three items on this list might well have led extraterrestrial observers to believe that, as a species, we displayed psychopathic and sociopathic tendencies. But it is
perhaps the last item on the list that might have caused the
most concern, because it would have been apparent to
them that this rocket technology represented the dawning
of the Space Age. This technology might eventually take
us and our weapons of mass destruction out into space—a
prospect that would almost certainly be viewed with
alarm. The extraterrestrials would have been correct in
their assessment, and indeed after the end of the Second
World War Wernher von Braun (who had developed the
V-2 rocket) became a naturalized American citizen and
used his skills to work on the technology that would eventually put a man on the moon. Apollo rockets were the direct descendants of the V-2.

Doubtless there is much on Earth that would impress
or intrigue our extraterrestrial observers, but the rapid advance of our technology coupled with our capacity for extreme violence may have convinced them that some action
was needed. Let's follow this thought up, and try to see
how we might appear to extraterrestrial observers. What
sort of a world do we *really* live in?

We live in a world where every day people die of
hunger while vast stocks of food are stored or even destroyed elsewhere. In November 1996 refugees in Zaire
ate grass in order to survive, while $80 million was spent

on a world summit on hunger. As the 6,000 delegates began to arrive at their hotels in Rome, an estimated 50,000 had already died in the Zairian tragedy, and over a million refugees were on the move. We live in a world where people die from preventable diseases; where every day people die in wars and conflicts fought over territorial disputes, politics or religion; where some regimes stay in power by such acts as using chemical weapons against their own citizens, or torturing or murdering dissenters; where some people go berserk for no apparent reason, and commit the most appalling crimes, such as the Dunblane massacre in Scotland, where sixteen primary school children and their teacher were shot dead, or the massacre in which Martin Bryant killed thirty-five people at a Tasmanian tourist resort. We live in a world where rape and the sexual abuse of children is widespread. We live in a world where chemical cocktails are put into the eyes of rabbits, so that yet another brand of perfume or shampoo can be launched. Racism and other such evils are commonplace.

I could go on, but I have made my point. I challenge every reader to buy a national newspaper today, and scan it. See the horror and misery that is reported, and think about how much never gets reported at all. Then reflect that you have simply studied the news for *one* day.

When we consider examples such as those mentioned above, is it not possible that extraterrestrial observers may be concerned at the threat we pose, and is it not likely that they would choose to do something about us? I believe that the abduction phenomenon could be a response to *our* behavior, and that it might represent a gradual attempt to effect subtle changes in humans. It is possible that the extraterrestrials are attempting to civilize us.

Such a hypothesis is not so much the result of listening to researchers, but stems from the far more valuable experience of listening to the abductees themselves. I would always expect to find more answers from those at the center of the phenomenon. The reason I favor this particular interpretation of events is based on the way in which abductees respond to their experiences.

I have repeatedly been told by people that they regard their experiences as having been transformative. One would expect the abductees to be furious with the aliens, but although many abductees are understandably angry, a surprising number have an emotional response that seems out of kilter with their experienced trauma. Their responses seem to be almost *artificial*. While researchers get angry, and talk about trying to prevent the abductions, the abductees themselves seek to forge a relationship with the aliens. They recall looking into the eyes of the leader being and feeling a sense of love, even during painful and invasive procedures. They leave books, poems and messages on their windowsills. They want to help with whatever is going on, and many have said that they will willingly assist the aliens in any way they can.

I believe that the central clue to the puzzle is the way in which many abductees have made major changes in their lives after their experiences. While I accept that from a psychological point of view *any* major event can have a profound effect on someone's life, there seems to be a definite pattern to the way in which abductees are affected.

We have seen that abductions often follow a standard pattern, and it looks as if this is equally true of the human response. The key point is the way in which certain individuals seem to respond, time and time again. Essentially, they become less ego-oriented, and more concerned with the plight of others. They develop psychic abilities and what appear to be spontaneous interests in spiritualism, they become more creative and particularly concerned with environmental issues. These are generalizations, and I should stress that these effects do not arise in all abductees. But they seem to arise in significantly greater numbers than chance might allow, especially when one considers that it is an apparently illogical and *inappropriate* response in the aftermath of what is often a painful and humiliating trauma.

It is the inappropriate nature of the response to the phenomenon that makes me think that this response has been *induced*. It seems to me that extraterrestrials, concerned at some of our worst excesses, might be carrying out a systematic program aimed at diverting our energies from violence, and placing them instead on more worthy pursuits. If humanity is to take its place among the stars, as our technological advances suggest that we will, it would not surprise me if our galactic neighbors viewed the prospect with alarm. They would surely wish to be certain that the appalling evil we have done on Earth is not to be repeated on a far wider scale.

The alien abduction phenomenon may be an attempt to change the course of human evolution. If this is the case, it would be nice to think that we might establish a dialogue with those responsible for this program, in order

to regain control of our own destiny, or at the very least
address our problems and failings in partnership with
these other intelligences. But perhaps the abduction phenomenon is a measure of last resort, taken by high-minded
beings who intervene reluctantly, and only because we
have steadfastly refused to curb our worst behavior. The
abduction phenomenon is regarded by some as a crime
against humanity. Perhaps it is designed to save us from
ourselves.

CONCLUSION

When I started to write this book my father asked me whether I knew what my conclusions were going to be. My father is a scientist, and I sensed that the question was one of those testing ones that parents like to ask. I said that although I had some views on the subject, I wouldn't be able to come to any conclusions until I'd seen where the data led. I think this was the answer he was hoping for, and the point is clear enough: it is unscientific to defend a personal view against all comers just to try to prove that you were right all along. You have to go with the flow, and see where it leads you. I recall seeing a television interview with the eminent cosmologist Stephen Hawking, in which he was asked about a statement he had made about time running backward. He had been putting forward some very exciting and exotic ideas, but he simply said that on reflection he felt that some of his previous ideas had been wrong. He had changed his mind. I was enormously impressed. One of the most important things that any researcher can learn from a great mind like Hawking's is that we are all fallible, and that nobody has all the answers.

I was initially skeptical about abductions. Even when it seemed to me that my official research at the Ministry pointed more and more to an extraterrestrial explanation for some UFOs, I remained distinctly uneasy about the concept of alien abduction. Like many researchers before me I felt that while the idea of alien craft visiting the Earth was acceptable, abductions were somehow beyond the pale. I shared the views of some of the early pioneers such as Keyhoe, thinking that abductions devalued ufology, and would serve to alienate not only serious researchers, but also the public, who would doubtless feel that a concept already difficult to accept had been hijacked by cranks and rendered unworthy of any serious study. But many investigators appreciate that, while there *are* undeniable cases of exaggeration within the field of abduction research, the opposite is also the case. In other words, some abductees will deliberately *understate* their experiences, feeling that this will make them more likely to be believed.

Slowly, my views on the phenomenon began to change. The sheer volume of reports, the commonality in independent accounts coupled with the physical and emotional effects on abductees convinced me that we were dealing with more than just hoaxes or psychological delusions. While such prosaic explanations undoubtedly accounted for *some* abductions, they could not account for them all.

Even after I became convinced of the physical reality of the alien abduction phenomenon, my views shifted and evolved. I started off feeling that genetic explanations, centering on the creation of a hybrid race, provided the answer to the mystery. I moved away from this idea in part because of the logical objections raised by researchers like Jacques Vallée, who was the first to ask why extraterrestrial visitors eager to obtain genetic material did not simply take it in bulk from laboratories, negating the need for the logistical nightmare that abductions must pose. Furthermore, I have a natural tendency to rebel against obvious solutions. I do not like to take the card that the magician wants me to take in a card trick. This may be little more than an unscientific hunch, but I cannot help but take it into account. But my main reason for moving away from such theories was because they did not generally tally with the views of the abductees I met, many of whom had a very positive feeling about their experiences. The apparent inappropriateness of their emotional responses, coupled with the frequent and spontaneous development of interests in issues such as spiritualism and environmental issues, suggested to me that other, more subtle factors were involved.

I accept that I may be wrong. Although I believe the evidence for the physical reality of abductions is so conclusive as to be virtually undeniable, I am the first to admit that my suggestion that the phenomenon is part of a campaign to civilize us as a species is only one more theory to join the ranks of competing ideas. It may not be the correct one, but this is comparatively unimportant. What is important is that the subject is brought to a wider audience, and this is why I have written this book.

At times this has been a difficult book to write. The distress experienced by some of the abductees has been sobering, and has acted as a brake on the idea that my theory may represent some sort of New Age interpretation of the phenomenon. It would be nice to think that the motivation behind abductions is to prepare us for entry into some sort of galactic network, but if this were the case, why would the extraterrestrials not be open with us? The covert nature of what is going on suggests that any alterations that are being made in human behavior are being carried out for the benefit of the abductors, and not for us. It is, perhaps, about removing the threat that we pose to them.

Putting this account together has sometimes been a confusing affair, as I grappled with complex and often contradictory data. It is impossible to avoid the wider debate about regression hypnosis and False Memory Syndrome, which is as controversial in relation to child sex abuse as it is to alien abductions. I reiterate my view that hypnotic memories should be not be treated as gospel—but then neither should consciously recalled memories. While hypnosis may not be the perfect investigative technique, it has provided corroboration with consciously recalled data. Whatever the problems with regression hypnosis, the challenge, perhaps, is to find a viable alternative.

Countless questions have been thrown up by my research into this phenomenon. In what reality or realities do these events occur? If the abductors use hypnosis or mind control to erase memories of the experiences, why is this not a complete process? Why do some memories remain? Are they left as tantalizing clues for us to follow? If so, what happens if public awareness of the abduction phenomenon grows to such an extent that the abductors regard erasing the memories as being wasted effort? How would we cope, as a species, with total conscious recall of widespread alien abductions?

This question of the true extent of the phenomenon interests me, and I believe that abductions may be far more common than many suspect. Some of these events may be consciously recalled while some may not be. Whichever, the abductees are aware of events at some level, and this causes stress. Even those who view their experiences in a positive light experience difficulties stemming from the fact that society does not generally acknowledge these experiences as real. Outside the UFO lobby, awareness of the phenomenon is limited, and is usually based on false assumptions created by trivialization of the subject by the media, and exaggeration by a handful of cranks. It is no surprise that people tend to keep these experiences to themselves, and I have lost count of the number of abductees I have talked to who have indicated that they had never previously reported their encounters to anyone.

There have been many fascinating stories that for a variety of reasons I have been unable to feature in this book. There is the celebrity who has told me of an experience containing a number of features frequently reported in abductions. Then there was the chance encounter in a restaurant with somebody who after a few drinks and a bit of chat told me an abduction story. What are the odds against walking into a public place and finding an abductee at the next table? Some might call this coincidence, but all my training and experience at the Ministry of Defence has taught me not to believe in coincidences. To me it is indicative of the fact that these experiences are incredibly common.

The most difficult dilemma that I had when writing this book was deciding how to handle the material I have about abductions involving children. Aside from the childhood experiences of those abductees who were happy to discuss them, I have generally kept such material out of the book. But I must mention two such cases, in order to give a picture of what is going on. I hope readers will forgive the almost total lack of personal details, and understand that witness confidentiality in such sensitive cases is paramount. I should also stress that I do not, on principle, quiz children about their experiences. Such a course of action would be intrusive and potentially harmful. The material in these cases came to me via the parents. I would advise extreme caution in cases involving children. Questioning should only be carried out by an appropriate expert (not a ufologist), and then, I would suggest, only if there is evidence that the child is suffering physical or psychological harm. Investigation for purely academic reasons is not to be recommended when children are involved.

I have one case where a young child reports being on a

craft, and playing with the aliens (perhaps some children's imaginary friends aren't imaginary after all). The child writes:

"There's ships that are very different in the sky. We think that they're real, then I feel very strange."

There are drawings to accompany this quote, and they show humanoid creatures with distinctive slanted, almond-shaped eyes. Sometimes they are pictured in an oval room, with computer screens. Sometimes they are by what looks like a UFO, and sometimes the child concerned is with the strange creatures.

A second incident involved a child who was with a babysitter. There was a choice of rooms to sleep in; the downstairs bedroom or the playroom at the top of the stairs. This room, with all the toys, might have seemed the natural choice, but the child was adamant:

"I don't want to sleep in that room. If I do, the bad fairies will come and take me away."

What of the psychological effects of the abduction experience? If the stresses that stem from bottling up those experiences that are viewed in a positive light can be traumatic, imagine what happens to the many people who view their experiences as negative—and do not let my theory downplay the undoubted fear, pain and anger that many people suffer as a direct result of an abduction. Whatever the solution to the mystery (if indeed there is a single neat solution), the abduction phenomenon is a very real experience that demands a response on four different levels.

The first response demanded by the phenomenon is a personal one. All abductees have to deal with their experiences, either by themselves or with the help of others. There are, essentially, two basic responses to abduction, these being either a positive one or a negative one. Given that there seems to be little one can do personally to prevent the experience, it makes more sense to try to come to a form of acceptance. As we have seen, research suggests that the witness is instrumental in interpreting and even shaping the phenomenon, which tends to mirror the belief system of the individual concerned. Positive thinking can therefore be important in helping people come to terms with their experiences.

The second response is that demanded from those ufologists who research and investigate abductions. These people play a key role in interpreting the phenomenon, and their belief systems can shape the experience in the same way as those of the abductees themselves. If anything, their influence is greater, because those who experience abduction look to these people for answers, and regarding them as experts, will give their views considerable weight. I urge that researchers do not abuse this influence. I make a plea for circumspection in handling investigations. This means extreme caution with hypnosis, only using it if you are qualified to do so. It means being open-minded enough to realize that your caseload will contain many examples of prosaic explanations. It means perhaps enrolling in a counseling course, and not being afraid to seek the advice of an appropriately qualified mental health professional. Above all, it means placing your own curiosity, ego, and understandable desire to uncover the case of the century in a secondary position to the needs of the abductee. We would all like to solve the mystery, but in some cases it may be in the best interests of the abductee not to proceed with an investigation at all.

The third response is demanded from the world of science. This is a phenomenon that is crying out for serious investigation. Ufologists have made a start, but mistakes have been made, and a handful of cranks and other irresponsible individuals have muddied the waters. Ufologists cannot and should not be doing this without assistance from mainstream scientists, whom I would urge to look at the matter with an open mind. There are exciting discoveries to be made here, but like so much of pioneering science, these may require innovative thinking that involves a movement away from traditional methodology, and into unfamiliar territory at the outer limits of our understanding.

The final response that the phenomenon demands is one from governments. Despite the best intentions of individuals, a true understanding of the abduction phenomenon will almost certainly need action to be coordinated and carried out by the state. Governments have the resources to undertake the in-depth study that we so desperately need, and the authority to lead and encourage a proper effort to discover and perhaps counter what is happening. The real crime against humanity that stems from alien abduction may not be that which is inflicted upon us by an extraterrestrial intelligence, but rather the indifference that allows the phenomenon to continue unchallenged.

REPORTING AN ABDUCTION

The organizations listed in Appendix 4 will be able to give help and advice on alien abductions.

Alternatively, anyone who believes they have had experiences similar to those described in this book is welcome to write to me, at the following address:

Nick Pope
c/o Simon & Schuster Ltd
West Garden Place
Kendal Street
London, England
W2 2AQ

All correspondence will be treated in the strictest confidence.

APPENDICES

31 August 1996

Dear Sir/Madam

I would be most grateful if you would let me have a statement detailing the policy and view of the Ministry of Defence in relation to the alien abduction phenomenon.

I assume that if any allegations are made by UK citizens that they have suffered harm as a result of actions carried out by occupants of unidentified craft that have penetrated

British airspace without authority, that this is a matter of concern to the Ministry of Defence; but if such allegations are the responsibility of another Government Department or agency, perhaps you would be kind enough to point me in the right direction.

Yours faithfully,

APPENDIX 2

FROM: **Secretariat (Air Staff) 2A1A, Room 8245**
Ministry of Defence
Main Building, Whitehall, London SW1A 2HB

1. Thank you for your letter dated 31 August 1996 regarding alien abduction. This office is the focal point within the Ministry of Defence for correspondence relating to 'unexplained' aerial sightings or 'UFOs'. I have been asked to reply.

2. First perhaps it would be useful if I were to explain the

role that the Ministry of Defence has with respect to 'unexplained' aerial sightings. The MOD examines any reports of 'UFO' sightings it receives solely to establish whether what was seen might have some defence significance; namely is there any evidence that the UK Air Defence Region might have been compromised by a foreign hostile military aircraft? The reports are examined with the assistance of the Department's air defence experts as required. Unless there is evidence of a potential military threat, and to date no 'UFO' sighting has revealed such evidence, we do not attempt to identify the precise nature of each sighting reported to us. We believe that down-to-earth explanations are available for most of these reported sightings, such as aircraft seen from unusual angles or natural phenomena.

3. You have asked for the MOD's policy and view in relation to alien abduction. Abduction is a criminal offence and as such is a matter for the civil police. As the MOD is not aware of any evidence which might substantiate the existence of extraterrestrial activity, the matter of abduction by alien lifeforms is a non-issue as far as the MOD is concerned.

4. I hope this explains our position.

Yours sincerely,

APPENDIX 3[1]

12 September 1996

Dear Sir/Madam

I am writing to ask whether the Home Office has a policy on the issue of alien abduction. I wrote to the Ministry of Defence, who said that as abduction was a criminal office, this was a matter for the police. I have attached a copy of their letter.

[1]No reply was received

Given that there are a number of UK citizens who have publicly reported being abducted by aliens, I would be interested to know if the Home Office or the police have taken any action.

Yours faithfully,

APPENDIX 4

Organizations Researching Alien Abductions

The following organizations are involved to varying degrees in researching and investigating the alien abduction phenomenon:

Academy of Clinical Close Encounter Therapists
2826 O Street
Suite 3
Sacramento
CA 95816

British UFO Research Association
BM BUFORA
London, England
WC1N 3XX

Center for UFO Studies
2457 West Peterson Avenue
Chicago
IL 60659

Fund for UFO Research
PO Box 277
Mt. Rainier
MD 20712

Intruders Foundation
PO Box 30233
New York
NY 10011

Mutual UFO Network
103 Oldtowne Road
Sequin
TX 78155-4099

Program for Extraordinary Experience Research
1493 Cambridge Street
Cambridge
MA 02139

UFO Magazine
Wharfebank House
Wharfebank Business Centre
Ilkley Road
Otley near Leeds, England
LS21 3JP

BIBLIOGRAPHY

Adamski, George: *Inside the Spaceships*. New York: Abelard-Schuman, 1955.

Adamski, George: *Flying Saucers Farewell*. New York: Abelard-Schuman, 1961.

Angelucci, Orfeo: *The Secret of the Saucers*. Amherst, WI: Amherst Press, 1955.

Bethurum, Truman: *Aboard a Flying Saucer*. Los Angeles: De Vorss and Company, 1954.

Bryan, C. D. B.: *Close Encounters of the Fourth Kind: Alien*

Abduction and UFOs—Witnesses and Scientists Report. London: Weidenfeld & Nicolson, 1995.

Budden, Albert: *UFOs: Psychic Close Encounters.* London: Blandford, 1995.

Bullard, Thomas E.: *UFO Abductions: The Measure of a Mystery.* Mt. Rainier, MD: The Fund For UFO Research, 1987.

Cassirer, Manfred: *Dimensions of Enchantment: The Mystery of UFO Abductions, Close Encounters and Aliens.* London: Breese Books, 1994.

Davenport, Marc: *Visitors From Time: The Secret of the UFOs.* Murfreesboro, TN: Greenleaf Publications, 1992.

Fiore, Edith: *Encounters: A Psychologist Reveals Case Studies of Abductions by Extraterrestrials.* New York: Ballantine Books, 1990.

Fowler, Raymond E.: *The Andreasson Affair.* Englewood Cliffs, NJ: Prentice-Hall, 1979.

Fry, Daniel W.: *The White Sands Incident.* Los Angeles: New Age Publishing Company, 1954.

Fuller, John G.: *The Interrupted Journey.* New York: Dial Press, 1966.

Good, Timothy, and Zinsstag, Lou: *George Adamski: The Untold Story.* London: Ceti Publications, 1983.

Good, Timothy: *Beyond Top Secret: The Worldwide UFO Security Threat.* London: Sidgwick & Jackson, 1996.

Hill, Betty: *A Common Sense Approach to UFOs.* Greenland, NH: Betty Hill, 1995.

Hopkins, Budd: *Missing Time: A Documented Study of UFO Abductions*. New York: Richard Marek Publishers, 1981.

Hopkins, Budd: *Intruders: The Incredible Visitations at Copley Woods*. New York: Random House, 1987.

Hopkins, Budd: *Witnessed: The True Story of the Brooklyn Bridge UFO Abductions*. New York: Pocket Books, 1996.

Howe, Linda Moulton: *An Alien Harvest: Further Evidence Linking Animal Mutilations and Human Abductions to Alien Life Forms*. Huntington Valley, PA: Linda Moulton Howe Productions, 1989.

Hynek, J. Allen: *The UFO Experience: A Scientific Inquiry*. Chicago: Henry Regnery Company, 1972.

Jacobs, David M.: *Secret Life: Firsthand Accounts of UFO Abductions*. New York: Simon & Schuster, 1992.

Jung, Carl J.: *Flying Saucers: A Modern Myth of Things Seen in the Sky*. New York: New American Library, 1959.

Keyhoe, Donald E.: *Flying Saucers From Outer Space*. New York: Henry Holt, 1953.

Keyhoe, Donald E.: *Aliens From Space*. Garden City, NY: Doubleday and Company, 1973.

Klass, Philip J.: *UFO Abductions: A Dangerous Game*. Buffalo: Prometheus Press, 1989.

Leslie, Desmond and Adamski, George: *Flying Saucers Have Landed*. London: Futura, 1953.

Lindemann, Michael (ed.): *UFOs and the Alien Presence: Six Viewpoints*. Santa Barbara, CA: The 2020 Group, 1991.

Lorenzen, Coral and Jim: *Abducted: Close Encounters of a Fourth Kind.* New York: Berkley Books, 1977.

Mack, John E.: *Abduction: Human Encounters with Aliens.* New York: Simon & Schuster, 1994.

Menger, Howard: *From Outer Space to You.* Clarksburg, WV: Saucerian Press, 1959.

Montgomery, Ruth: *Strangers Among Us.* New York: Fawcett Crest, 1979.

Montgomery, Ruth: *Aliens Among Us.* New York: Fawcett Crest, 1985.

Nagaitis, Carl and Mantle, Philip: *Without Consent: A Comprehensive Survey of Missing-Time and Abduction Phenomena in the UK.* London: Ringpull, 1994.

Oakensen, Elsie: *One Step Beyond . . . : A Personal UFO Abduction Experience.* London: Regency Press, 1995.

Pope, Nick: *Open Skies, Closed Minds: For the First Time a Government UFO Expert Speaks Out.* London: Simon & Schuster, 1996.

Pritchard, Andrea et al (ed.): *Alien Discussions: Proceedings of the Abduction Study Conference.* Cambridge, MA: North Cambridge Press, 1994.

Randles, Jenny: *Abduction: Over 200 Documented UFO Kidnappings.* London: Robert Hale, 1988.

Rimmer, John: *The Evidence for Alien Abductions.* Wellingborough, Northamptonshire: Aquarian Press, 1984.

Rogo, D. Scott (ed.): *UFO Abductions: True Cases of Alien Kidnappings.* New York: Signet, 1980.

Ruppelt, Edward J.: *The Report on Unidentified Flying Objects*. Garden City, NY: Doubleday and Company, 1956.

Sagan, Carl: *The Demon-Haunted World: Science as a Candle in the Dark*. London: Headline, 1996.

Schnabel, Jim: *Dark White: Aliens, Abductions and the UFO Obsession*. London: Hamish Hamilton, 1994.

Spencer, John: *Perspectives: A Radical Examination of the Alien Abduction Phenomenon*. London: Macdonald & Co. (Publishers) Ltd., 1989.

Steiger, Brad: *The UFO Abductors*. New York: Berkley Books, 1988.

Strieber, Whitley: *Communion: A True Story*. New York: William Morrow & Co., 1987.

Strieber, Whitley: *Transformation: The Breakthrough*. New York: William Morrow & Co., 1988.

Strieber, Whitley: *Breakthrough: The Next Step*. London: Simon & Schuster, August 1997.

Strieber, Whitley: *The Secret School*. London: Simon & Schuster, August 1997.

Vallee, Jacques: *Passport to Magonia*. Chicago: Henry Regnery, 1969.

Vallee, Jacques: *Dimensions: A Casebook of Alien Contact*. New York: Ballantine Books, 1988.

Vallee, Jacques: *Confrontations: A Scientist's Search for Alien Contact*. New York: Ballantine Books, 1990.

Van Tassel, George W.: *I Rode a Flying Saucer.* Los Angeles: New Age Publishing Company, 1952.

Walters, Ed and Frances: *The Gulf Breeze Sightings.* New York: William Morrow & Co., 1990.

Walton, Travis: *The Walton Experience.* New York: Berkley Books, 1978.

INDEX